A Student's Guide to Data and Error Analysis

All students taking laboratory courses within the physical sciences and engineering will benefit from this book, whilst researchers will find it an invaluable reference. This concise, practical guide brings the reader up to speed on the proper handling and presentation of scientific data and its inaccuracies. It covers all the vital topics with practical guidelines, computer programs (in Python), and recipes for handling experimental errors and reporting experimental data. In addition to the essentials, it also provides further background material for advanced readers who want to understand how the methods work. Plenty of examples, exercises, and solutions are provided to aid and test understanding, whilst useful data, tables, and formulas are compiled in a handy section for easy reference.

HERMAN J. C. BERENDSEN is Emeritus Professor of Physical Chemistry at the University of Groningen, the Netherlands. His research started in nuclear magnetic resonance, but focused later on molecular dynamics simulations on systems of biological interest. He is one of the pioneers in this field and, with over 37 000 citations, is one of the most quoted authors in physics and chemistry. He has taught courses in molecular modeling worldwide and authored the book *Simulating the Physical World* (Cambridge University Press, 2007).

A Student's Guide to
Data and Error Analysis

HERMAN J. C. BERENDSEN

Emeritus Professor of Physical Chemistry,
University of Groningen, the Netherlands

CAMBRIDGE
UNIVERSITY PRESS

CAMBRIDGE UNIVERSITY PRESS
Cambridge, New York, Melbourne, Madrid, Cape Town,
Singapore, São Paulo, Delhi, Mexico City

Cambridge University Press
The Edinburgh Building, Cambridge CB2 8RU, UK

Published in the United States of America by Cambridge University Press, New York

www.cambridge.org
Information on this title: www.cambridge.org/9780521119405

© H. Berendsen 2011

First published 2011

A catalogue record for this publication is available from the British Library

Library of Congress Cataloguing in Publication Data

Berendsen, Herman J.C.
A student's guide to data and error analysis / Herman J.C. Berendsen.
p.   cm.
ISBN 978-0-521-11940-5 (Hardback) – ISBN 978-0-521-13492-7 (pbk.)
1. Error analysis (Mathematics)   I. Title.
QA275.B43 2011
511'.43–dc22

2010048231

ISBN 978-0-521-11940-5 Hardback
ISBN 978-0-521-13492-7 Paperback

To my wife and daughters

Contents

Preface

This book is written as a guide for the presentation of experimental data including a consistent treatment of experimental errors and inaccuracies. It is meant for experimentalists in physics, astronomy, chemistry, life sciences and engineering. However, it can be equally useful for theoreticians who produce simulation data: they are often confronted with statistical data analysis for which the same methods apply as for the analysis of experimental data. The emphasis in this book is on the determination of best estimates for the values and inaccuracies of parameters in a theory, given experimental data. This is the problem area encountered by most physical scientists and engineers. The problem area of experimental design and hypothesis testing – excellently covered by many textbooks – is only touched on but not treated in this book.

The text can be used in education on error analysis, either in conjunction with experimental classes or in separate courses on data analysis and presentation. It is written in such a way – by including examples and exercises – that most students will be able to acquire the necessary knowledge from self study as well. The book is also meant to be kept for later reference in practical applications. For this purpose a set of "data sheets" and a number of useful computer programs are included.

This book consists of parts. Part I contains the main body of the text. It treats the most common statistical distributions for experimental errors and emphasizes the error processing needed to arrive at a correct evaluation of the accuracy of a reported result. It also pays attention to the correct reporting of physical data with their units. The last chapter considers the inference of knowledge from data from a Bayesian point of view, hopefully inducing the reader to sit back and think. The material in Part I is kept practical, without much discussion of the theoretical background on which the various types of analysis are based. This will not at all satisfy the eager student who has sufficient background in mathematics and who wishes to grasp a fuller understanding of the principles involved. Part II is to satisfy the curious: it contains several Appendices that explain various issues in more detail and provide derivations of the equations quoted in Part I. The Appendices in

Part II obviously require more mathematical skills (in particular in the field of linear algebra) than Part I. Part III contains Python code examples and Part IV provides answers to exercises. Finally, Part V contains practical information in the form of a number of "data sheets" which provide reference data in a compact form.

Throughout the book computer programs are included to facilitate the computations needed for applications. There are several professional software packages available for statistical data analysis. In the context of an educational effort, I strongly advise against the use of a specialized "black-box" software package that can be easily misused to produce ill-understood results. A "black-box" computer program should never be a magic substitute for a method that is not understood by the user! If a software package is to be used, it should provide general mathematical and graphical tools, preferably in an interactive way using an interpreter rather than a compiler. The commercial packages MATHEMATICA, MATLAB and MATHCAD are suitable for this purpose. However, most readers of this book will not have access to any or all of these packages, or – if they have temporary access through their institution – may not be able to continue access at a later point in time. Therefore for this book the choice was made to use the generally available, actively developing, open-source interpretative language PYTHON. With its array-handling and scientific extensions NUMPY and SCIPY the capabilities of this language come close to those of the commercial packages. Software related to this book, including a Python module `plotsvg.py` providing easy plotting routines, can be found on www.hjcb.nl/.

This book is the successor of the Dutch textbook *Goed meten met fouten* (Berendsen, 1997) that has been used in courses at the departments of physics and chemistry of the University of Groningen since 1997. The author is indebted to Emile Apol, A. van der Pol and Ruud Scheek for corrections and suggestions. Comments from readers are welcome to author@hjcb.nl.

PART I

Data and error analysis

1 Introduction

It is impossible to measure physical quantities without errors. In most cases errors result from deviations and inaccuracies caused by the measuring apparatus or from the inaccurate reading of the displaying device, but also with optimal instruments and digital displays there are always fluctuations in the measured data. Ultimately there is random thermal noise affecting all quantities that are determined at a finite temperature. Any experimentally determined quantity therefore has a certain inaccuracy. If the experiment were to be repeated, the result would be (slightly) different. One could say that the result of a particular experiment is no more than a *random sample* from a probability distribution. When reporting the result of an experiment, it is important to also report the extent of the uncertainty, e.g. in terms of the best estimate of some measure of the *width* of the probability distribution. When experimental data are processed and conclusions are drawn from them, knowledge of the experimental uncertainties is essential to assess the reliability of the conclusion.

Ideally, you should specify the probability distribution from which the reported experimental value is supposed to be a random sample. The problem is that you have only one experiment; even if your experiment consists of many observations of which you report the average, you have only one average to report. So you have only one sample of the reported item and you could naively conclude that you have no knowledge at all about the underlying probability distribution of that sample. Fortunately, there is the science of statistics that tells us differently. When your experiment consists of a series of repeated observations of a variable x, with outcomes $x_1, x_2, \ldots, x_n$, and you report the result of the total experiment as the average of the x_i's, statistics tells you how to *estimate* certain properties of the probability distribution of which the reported result is supposed to be a random sample. Thus you can estimate the mean of the distribution or – if you prefer – the most probable value of the distribution, which then is the result of your measurement. You can also estimate the width of the distribution, which indicates the random uncertainty in the result.

The result of an experiment is generally not equal to a directly measured quantity, but is derived from measured quantities by some functional relation.

3

For example, the area of a rectangle is the product of the measured length and width of two sides. Each measurement has its estimated value and random error and these errors *propagate* through the functional relation (here a product) to the final result. The contributing errors must be properly combined to one error estimate in the result.

The purpose of this book is to indicate how one can arrive at the best estimates of both the value(s) and the random error(s) in the result, based on the measurements from which the result is derived. In order to maintain its usefulness as a practical guide, the main part of this book simply states the equations and procedures, without proper derivations. Thus the practical applicant is not bothered by unnecessary detail. However, several appendices are included that provide further details and give a proper background in statistics with derivations of the equations used. For further reading many textbooks are available.[1]

Chapter 2 describes the proper presentation of results of measurements with their accuracies and with their units. Chapter 3 classifies the various types of error and describes how contributing errors will propagate and combine into a more complex result. Chapter 4 describes a number of common probability distributions from which experimental errors may be sampled. In Chapter 5 it is shown how the characteristics of a *data series* can be defined and then be used to arrive at estimates of the best value and accuracy of the result. Chapter 6 is concerned with simple graphic treatment of data, while Chapter 7 treats the more accurate *least-squares* fitting of model parameters to experimental data. Chapter 8, finally, discusses the philosophical basis of statistical methods, confronting traditional hypothesis testing with the more intuitive but powerful *Bayesian* method to determine the probability distribution of model parameters.

---

[1] Most textbooks aim at a wider audience and are therefore less useful for physical scientists and engineers. For the latter interest group see Bevington and Robinson (2003), Taylor (1997), Barlow (1989) and Petruccelli *et al.* (1999).

2 The presentation of physical quantities with their inaccuracies

This chapter is about the *presentation* of experimental results. When the value of a physical quantity is reported, the uncertainty in the value must be properly reported too, and it must be clear to the reader what kind of uncertainty is meant and how it has been estimated. Given the uncertainty, the value must be reported with the proper number of digits. But the quantity also has a unit that must be reported according to international standards. Thus this chapter is about reporting your results: this is the last thing you do, but we'll make it the first chapter before more serious matters require attention.

2.1 How to report a series of measurements

In most cases you derive a result on the basis of a series of (similar) measurements. In general you do not report all individual outcomes of the measurements, but you report the best estimates of the quantity you wish to "measure," based on the experimental data and on the model you use to derive the required quantity from the data. In fact, you use a *data reduction method*. In a publication you are *required* to be explicit about the method used to derive the end result from the data. However, in certain cases you may also choose to report details of the data themselves (preferably in an appendix or deposited as "additional material"); this enables the reader to check your results or apply alternative data reduction methods.

List all data, a histogram or percentiles

The fullest report of your experimental data is a list or table of all data. Almost[1] equivalent is the report of a *cumulative distribution* of the data (see Section 5.1 on page 54). Somewhat less complete is reporting a *histogram* after collecting data in a limited number of intervals, called *bins*. Much less

---

[1] Not quite, because one loses information on possible sequential correlation between data points.

Table 2.1 *Thirty observations, numbered in increasing order.*

1	6.61	6	7.70	11	8.35	16	8.67	21	9.17	26	9.75
2	7.19	7	7.78	12	8.49	17	9.00	22	9.38	27	10.06
3	7.22	8	7.79	13	8.61	18	9.08	23	9.64	28	10.09
4	7.29	9	8.10	14	8.62	19	9.15	24	9.70	29	11.28
5	7.55	10	8.19	15	8.65	20	9.16	25	9.72	30	11.39

complete is to report certain *percentiles* of the cumulative distribution, usu-ally the 0, 25%, 50%, 75% and 100% values (i.e., the full range, the median and the first and third quartiles). This is done in a *box-and-whisker* display. See the example below.

List properties of the data set

The methods above are *rank-based* reports: they follow from ranking the data in a sequence. You can also report *properties* of the set of data, such as the number of observations, their average, the mean squared deviation from the average or the root of that number, the correlation between successive observations, possible outliers, etc. Note that we do not use the names *mean, variance, standard deviation*, which we reserve for properties of probabil-ity distributions, not data sets. Use of these terms may cause confusion; for example, the *best estimate* for the variance of the parent probability distri-bution – of which the data set is supposed to be a random sample – is not equal to the mean squared deviation from the average, but slightly larger $(n/(n-1)\times)$. See Section 5.3 on page 58.

Example: 30 observations

Suppose you measure a quantity x and you have observed 30 samples with the results as given in Table 2.1. Figure 2.1 shows the cumulative distribution function of these data and Fig. 2.2 shows the same, but plotted on a "proba-bility scale" which should produce a straight line for normal-distributed data. A histogram using six equidistant bins is shown in Fig. 2.3. It is clear that this sampling is rather unevenly distributed.

These numbers and cumulative distributions were generated with **Python code** 2.1 on page 171

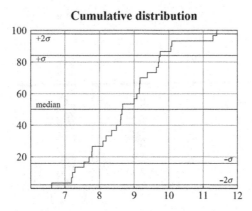

Figure 2.1 The cumulative distribution function of thirty observations. The vertical scale represents the cumulative percentage of the total.

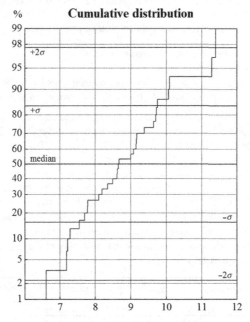

Figure 2.2 The cumulative distribution function of thirty observations. The vertical scale represents the cumulative percentage of the total on a probability scale, designed to produce straight lines for normal distributions.

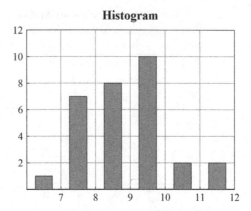

Figure 2.3 A histogram of thirty observations. The data have been gathered in six equidistant bins. The vertical scale gives the number of observations in each bin.

 The histogram of Fig. 2.3 was generated with **Python code** 2.2 on page 171

The *properties* of the data set you can report are:

(i)   *number of observations*: $n = 30$
(ii)   *average*: $m = 8.78$
(iii) *mean squared deviation from average*: msd $= 1.28$
(iv) *root-mean-square deviation from average*: rmsd $= 1.13$

 The properties are available as array methods or functions. See **Python code** 2.3 on page 171

Other rank-based properties of the data set are values that exceed a given fraction of the data, such as the *median* (at 50%), the first and third *quartile* (at 25 and 75%) or the p-th *percentile*. The latter is a value x_p such that $p\%$ of the data has a value $\leq x_p$ and $(100 - p)\%$ has a value $> x_p$.[2] The total *range* is the interval between the minimum and maximum values. Figure 2.4 shows the data as a box-and-whisker display of the total range (the whisker) and the quartiles (the box).

 A simple program to determine a series of percentiles is **Python code** 2.4 on page 172

---

[2] There may be an ambiguity here. The p-th percentile may be exactly one of the data values, e.g. the median equals the 5th value out of a set of 9. In general, the percentile will fall in a range between two values, e.g. the median lies between the 5th and the 6th value out of a set of 10 values. In that case linear interpolation is used.

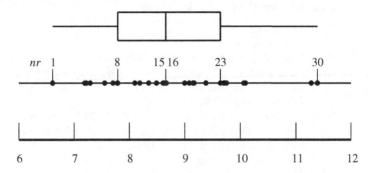

Figure 2.4 A *box-and-whisker* display of the total range, the median and the first and third quartile of thirty observations. Note that the median falls between nr 15 and nr 16 (15 observations on both sides); the average between the two values is taken.

2.2 How to represent numbers

Decimal separator: comma or period?

In the English language and in all "computer languages" (and among others also in China, Israel and Switzerland) the decimal *point* is used as separator between the integer and fractional parts of a real decimal number. In many other languages (all other European languages, Russian and related languages) the decimal *comma* is used instead. Be consistent and adhere to what your language requires! In order to avoid confusion, scientists are strongly advised *not* to use periods or commas to divide long numbers into groups of three digits, like 300,000 (English) or 300.000 (e.g. French). Instead, use a *space* (or even better, if your text editor allows it, a *thin space*) to separate groups of three digits: 300 000.[3]

Significant figures

The end result of a measurement must be presented with as many digits as are compatible with the accuracy of the result. Also when a number ends with zeros! These are the *significant figures* of the result. However, intermediate results in a calculation should be expressed with a higher precision in order to prevent accumulation of rounding errors. Always indicate the accuracy of the end result! If the accuracy is not explicitly given, it is assumed that the error in the last digit is ±0.5.

---

[3] This is the IUPAC recommendation, see http://old.iupac.org/reports/provisional/guidelines. html#printing

Examples, for the English language

(i) 1.65 ± 0.05

(ii) 2.500 ± 0.003

(iii) $35\,600 \pm 200$: better as $(3.56 \pm 0.02) \times 10^4$

(iv) 5.627 ± 0.036 is allowed, but makes sense only when the inaccuracy itself is known with sufficient accuracy. If not, this value should be written as 5.63 ± 0.04.

(v)   Avogadro's number is known as $(6.022\,141\,79 \pm 0.000\,000\,30) \times 10^{23}$ mol^{-1} (CODATA 2006). The notation $6.022\,141\,79(30) \times 10^{23} \text{ mol}^{-1}$ is a commonly accepted abbreviation.

(vi) 2.5 means 2.50 ± 0.05

(vii) 2.50 means 2.500 ± 0.005

(viii) In older literature one sometimes finds a subscript $_5$, indicating an inaccuracy of about one quarter in the last decimal: $2.3_5 = 2.35 \pm 0.03$, but this is not recommended.

When inaccuracies must be rounded, then do this in a conservative manner: when in doubt, round up rather than down. For example, if a statistical calculation yields an inaccuracy of 0.2476, then round this to 0.3 rather than 0.2, unless the statistics of your measurement warrants the expression in two decimals (0.25). See Section 5.5 on page 60. Be aware of the fact that calculators know nothing about statistics and generally suggest a totally unrealistic precision.

2.3 How to express inaccuracies

There are many ways to express the (in)accuracy of a result. When you report an inaccuracy it must be absolutely clear which kind of inaccuracy you mean. In general, when no further indication is given, it is assumed that the quoted number represents the *standard deviation* or *root-mean-square error* of the estimated probability distribution.

Absolute and relative errors

You can indicate inaccuracies as *absolute*, with the same dimension as the reported quantity, or as a dimensionless *relative* value. Absolute inaccuracies are often given as numbers in parentheses, relating to the last decimal(s) of the quantity itself.

Examples

(i) 2.52 ± 0.02

(ii) $2.52 \pm 1\%$

(iii) $2.52(2)$

(iv) $N_A = 6.022\,141\,79(30) \times 10^{23} \text{ mol}^{-1}$

Using probability distributions

If the degree of knowledge you have about the reported quantity θ can be expressed as a *probability distribution* of that quantity, you can report one or more *confidence intervals*. This is usually the case if a Bayesian analysis has been made (see Chapter 8). In cases where the estimated probability distribution deviates significantly from a Gaussian shape and the variance or standard deviation may be a meaningless or uninformative quantity, a confidence interval is the best way to report the accuracy of a quantity. One then gives the *Bayesian estimate* for the quantity as the expectation (mean) over the distribution, with e.g. a 90 percent confidence interval. That interval is given by two values; the probability that the quantity is less than the lower boundary is 0.05 and the probability that it exceeds the higher boundary is also 0.05. For the reader's information it is advised to also give the number n of independent experiments on which the estimate is based.

There are several possibilities to express the *estimated value* $\hat{\theta}$:

(i)   the *mean* or expectation over the probability distribution $E[\theta] = \int \theta \, p(\theta) \, d\theta$,

(ii) the *median*, i.e., the value for which the cumulative distribution (see Section 4.2 on page 29) reaches 50 percent. The probability that the real value is smaller than the median equals the probability that it is larger,

(iii) the *mode* or *most probable value*, which marks the maximum of the probability distribution.

These estimates are similar and in general their difference is insignificant, being much less than the standard deviation. For symmetric distributions they are all equal. In any case be explicit as to the kind of estimate you report.

Examples

(i)   In a simulation you "observe" the occurrence of a certain event (e.g. a conformational change of a protein molecule) that is irreversible on the attainable time scale. Your theory predicts that the event occurs with constant probability $k\Delta t$ in any small time interval Δt. You observe seven such events (occurring at $t_1, t_2, \ldots, t_7$) and apply a Bayesian analysis (see Chapter 8, page 120) to derive a probability distribution for k. The expectation of the rate constant is $E[k] = 7/(t_1 + t_2 + \cdots t_7) = 1.0 \, \text{ns}^{-1}$. This distribution $p(k)$, with cumulative distribution $P(k)$ (see Fig. 2.5) has the following properties (given in too many decimals):

- The *mean* equals 1.00; this is the best estimate $\hat{k}$.
- The *median* equals 0.95.
- The *mode* is 0.86.

Bayesian probability distribution

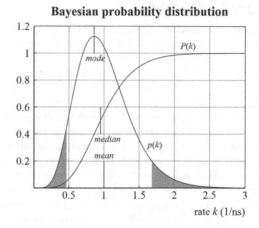

Figure 2.5 The Bayesian probability distribution for the rate k of an exponential decay process, based on seven lifetime observations.

- The *standard deviation*, i.e., the square root of the expectation of the squared deviation from the mean: $\hat{\sigma} = \sqrt{E[(k - \hat{k})^2]} = 0.38$. Just to see how well the standard deviation describes the width of the distribution: for a normal distribution 68% of the cumulative probability lies in the interval $(\hat{k} - \hat{\sigma}, \hat{k} + \hat{\sigma})$; for the Bayesian distribution of this example 69% is in the range $(1 - 0.38, 1 + 0.38)$. So the central region of the distribution is much like normal and the use of the standard deviation really makes sense in that range. But the tails are quite different!
- 90% confidence levels: $k(P = 0.05) = 0.47$; $k(P = 0.95) = 1.69$. This means that there is a 90% probability that the value of k lies between 0.5 and 1.7.

In this case you can report all "experimental" values $t_1, t_2, \ldots, t_7$, allowing the reader to draw her/his own conclusions. The result can be reported in various ways. The simplest is $\hat{k} = 1.0 \pm 0.4$, but that says nothing about the kind of distribution. It is better to give in addition a confidence interval and the number of observations, e.g.:

90% Bayes confidence interval $= (0.5, 1.7); n = 7$.

If you wish to be exhaustive, give the full probability distribution as in Fig. 2.5.

(ii) You measure the velocity of particles in a particle beam by time-of-flight determinations of 100 individual particles. Each velocity value is a sample from an unknown distribution. You want to determine two properties of the beam: the *mean* and the *standard deviation* of the one-dimensional (forward) velocity distribution of all particles in the beam. This kind of problem is treated in Section 5.2 on page 57. Your set of 100 measurements is characterized by an average $\langle v \rangle$ of 1053 m/s and a mean square deviation from the average $\langle (\Delta v)^2 \rangle$ of 2530 m^2/s^2. Here, $\Delta v = v - \langle v \rangle$. For each property you wish to give the best estimate and its standard error. You may report:

- mean velocity: 1053 ± 5 m/s,
- s.d. of velocity distribution: 50 ± 4 m/s.

  Realizing that you do not know beforehand what the variance of the distribution is, you may apply Student's t-distribution (see Section 5.4 on page 59) and report:

- mean velocity: 1053 m/s; 90% t-distr. confidence interval = (1045, 1061) m/s, $\nu = 99$.

In this case with a large number of degrees of freedom, reporting a t-distribution confidence interval is hardly meaningful, as the difference with a normal distribution is negligible. Reporting the standard error is much better.

2.4 Reporting units

SI units

Physical quantities not only have a numerical value with inaccuracy, but also a *unit*. *Always include the proper unit in the correct notation when you report a physical quantity.* There are international agreements on units and notation. The agreed system of units is the "Système International d'Unités" (SI).[4] The SI units are derived from the *SI base units* m, kg, s, A, K, mol, cd (see data sheet UNITS on page 215). You should make it a habit to adhere strictly to these units, even if you are often confronted with non-SI units in the literature (dominantly originating from the USA). So, kJ/mol and not kcal/mol, nm (or pm) and not Å, N and not kgf, Pa and not psi.

Non-SI units

Some non-SI units are allowed, such as the minute (min), hour (h), day (d), degree (°), minute angle ('), second angle ("), liter (L = dm^3), metric ton (t = 1000 kg)) and astronomical unit (ua = $1.495\,978\,70 \times 10^{11}$ m). Chemists use the liter a lot: note that its symbol is upper case L, *not* (as is still quite common) lower case l.[5] Thus a milliliter is mL, not ml. The concentration unit mol/L is allowed next to the SI unit mol/m^3, but the symbol M for *molar*

---

[4] The SI system was established in 1960 by the CGPM (Conférence Générale des Poids et Measures), an intergovernmental treaty organization.

[5] This is the CGPM recommendation since 1979.

(= mol/L) is now obsolete. Note that mol is written with lower case m and without an e as in *mole*. A *mole* is the English name for the quantity that has the unit mol. Some other non-SI units are not officially allowed, but often used within restricted contexts, such as the nautical mile (= 1852 m), knot (nautical mile/h), are (100 m^2), hectare (10^4 m^2), ångström (Å $= 10^{-10}$ m), barn ($b = 10^{-28}$ m^2) and bar (10^5 Pa). Use only official notations; not sec but s, not gr but g, not micron but μm. Use the *prefixes* as tabulated in the data sheet DATA:UNITS on page 215, preferably as powers that are multiples of 3. Be careful with capitalization: not m for Mega, not g for Giga. Finally, do not use confusing notations for more complex units, such as two slashes: not kg/m/s or kg/m s, but kg m^{-1} s^{-1}.

Typographical conventions

There are also agreed *typographical conventions*, which should be adhered to not only in scientific manuscripts, but even in informal reports. With modern text editors there is no excuse not to use roman, italic or bold type when required. The rules are simple:

- italic type for scalar quantities and variables,
- roman type for units and prefixes (mind capitalization),
- italic boldface for a vector or matrix quantity,
- sans-serif bold italic for tensors,
- roman type for chemical elements and other descriptive terms, including mathematical constants, functions and operators.

Examples

(i)   The input voltage $V_{in} = 25.2$ mV,
(ii) The molar volume $V_m = 22.4$ L/mol,
(iii) The force on the i-th particle $F_i = 15.5$ pN,
(iv) The symbol for nitrogen is N, the nitrogen molecule is N_2,
(v)   A nitrogen oxide mixture NO_x with $x = 1.8$,
(vi) $e = 2.718\ldots$; $\pi = 3.14\ldots$,
(vii) $F = ma = -$ **grad** V,
(viii) The surviving fraction of the k-th species, $f_k^{surv}(t) = \exp(-t/\tau_k)$.

2.5 Graphical presentation of experimental data

Experimental results are often presented in graphical form. The expectation or mean is given as the position (x, y) of (the center of) a symbol in a plot. The usual representation of inaccuracies in x and/or y is an *error bar* with a total length of twice the standard error. While both x and y values may be subject to experimental errors, very often one of the values (usually x)

Table 2.2 *Concentration of a reactant as a function of time. The inaccuracy is given as the estimated standard error.*

time t/s	conc. c/mmol L^{-1} $\pm$ s.d.
20	75 ± 4
40	43 ± 3
60	26 ± 3
80	16 ± 3
100	10 ± 2
120	5 ± 2
140	3.5 ± 1.0
160	1.8 ± 1.0
180	1.6 ± 1.0

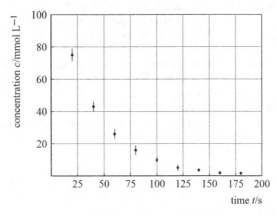

Figure 2.6 A linear plot of concentration of a reactant versus time, with error bars representing $\pm$ the standard error. The data are given in Table 2.2.

is so accurate that it makes no sense to plot an error bar. Figures 2.6 and 2.7 give examples of such a graphical representation, using the data given in Table 2.2. The reason to use a logarithmic scale for the concentration is that an expected exponential decay with time would show up as a straight line.

The linear plot can hardly show the small standard deviations of the last three points; on the logarithmic plot the s.d. on the small values show as much

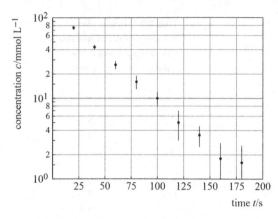

Figure 2.7 A logarithmic plot of the same data.

larger and asymmetric error bars. Note that the error bars on the last two points extend below the lower limit (1 mmol/L) of the logarithmic scale and therefore appear too short on the graph. Negative ordinate values (which may occur as a result of random errors) cannot be shown at all on a logarithmic scale.

Some authors use "whiskers" marking the ends of the error bars in order to make them more visible, but this does not add any useful information to the whiskerless error bars.

A scientifically acceptable graph should indicate *which* variables are plotted on the horizontal and vertical axes and *what units* the numbers represent. It is acceptable is to put the units in parentheses: time t (s), but it is advisable to use the notation time t/s as in Figs. 2.6 and 2.7; this notation indicates a dimensionless quantity represented by the numbers along the axes. Both notations are acceptable as long as the notation is consistent throughout a publication. Don't use more than one forward slash: $E_{pot}/kJ\,mol^{-1}$ is OK, but $E_{pot}/kJ/mol$ is not!

Of course there are several plotting programs to realize fancy graphics on a computer, but in many cases a quick sketch by hand on graph paper suffices to get a crude idea of a functional relationship and the importance of inaccuracies.

 See **Python code** 2.5 on page 172 to produce Fig. 2.7 with a logarithmic scale.

Summary   *In this chapter you have seen how experimental results are properly presented in a report or publication. A proper presentation gives numerical values with as many*

*significant digits as is warranted by the precision of the value,
it includes an unequivocal report of the (in)accuracy of the
value, it includes the proper unit in which the result is expressed
and it uses the conventional typography for variables, numbers,
units and prefixes. Experimental results are always presented
with an error estimate and it must be made explicit what the
reported inaccuracy means and how the error estimate has been
obtained.*

Exercises

2.1 Correct the following notations:

(a) $l = 3128 \pm 20$ cm,

(b) $c = 0.01532$ mol/L ± 0.1 mmol/L,

(c) $\kappa = 2.52 \times 10^2$ A m^{-2}/V m^{-1},

(d) k/L mol^{-1}/s $= 3571 \pm 2\%$,

(e) $g = 2 \pm 0.03$.

2.2 Convert the following quantities to SI units or units allowed within the SI system (see data sheet UNITS on page 215):

(a) a pressure of 1.30 mm Hg,

(b) a pressure of 33.5 psi,

(c) a concentration of 2.3 mM (millimolar),

(d) an interatomic distance of 1.45 Å,

(e) an activation energy of 5.73 kcal/mol,

(f) a daily energy requirement of 2000 calories,

(g) a force of 125 lbf,

(h) an (absorbed) radiation dose of 20 mrad,

(i) a fuel consumption of 3.4 (US) gallon per 100 mile,

(j) a dipole moment of 1.85 debye,

(k) a polarizability of 1.440 Å^3. Note that polarizability α in rationalized SI units is the ratio of induced dipole moment (in Cm) and electric field (in V/m). In unrationalized units the polarizability is $\alpha' = \alpha/(4\pi\varepsilon_0)$, expressed in units of volume.

3 *Errors: classification and propagation*

There are errors and uncertainties. The latter are unavoidable; eventually it is the omnipresent thermal noise that causes the results of measurements to be imprecise. After trying to identify and correct avoidable errors, this chapter will concentrate on the propagation and combination of uncertainties in composite functional relations.

3.1 Classification of errors

There are several types of error in experimental outcomes:

(i)   (accidental, stupid or intended) mistakes
(ii) systematic deviations
(iii) random errors or *uncertainties*

The first type we shall ignore. Accidental mistakes can be avoided by careful checking and double checking. Stupid mistakes are accidental errors that have been overlooked. Intended mistakes (e.g. selecting data that suit your purpose) purposely mislead the reader and belong to the category of *scientific crimes*.

Systematic errors

Systematic errors have a non-random character and distort the result of a measurement. They result from erroneous calibration or just from a lack of proper calibration of a measuring instrument, from careless measurements (uncorrected parallax, uncorrected zero-point deviations, time measurements uncorrected for reaction time, etc.), from impurities in materials, or from causes the experimenter is not aware of. The latter are certainly the most dangerous type of error; such errors are likely to show up when results are compared to those of other experimentalists at other laboratories. Therefore independent corroboration of experimental results is required before a critical experiment (e.g. one that overthrows an accepted theory) can be trusted.

18

Random errors or uncertainties

Random errors are unpredictable by their very nature. They can be caused by the limited precision of instrumental readings, but are ultimately due to *physical noise*, i.e. by natural fluctuations due to thermal motions or to the random timing of single events. Since such errors are unavoidable and unpredictable, the word "error" does not convey the proper meaning and we prefer to use the term *uncertainty* for the possible random deviation of a measured result from its true value.

If a measurement is repeated many times, the results will show a certain *spread* around an average value, from which the estimated inaccuracy in the average can be determined. The probability distribution, from which the measured values are *random samples*, is supposed to obey certain statistical relations, from which rules to process the uncertainties can be derived. In the case of a single measurement one should estimate the uncertainty, based on knowledge of the measuring instrument. For example, a length read on a ruler will be accurate to ± 0.2 mm; a length read on a vernier caliper will be accurate to ± 0.05 mm. Chemists reading a liquid level on a buret or graduated cylinder can estimate volumes with a precision of ± 0.3 scale divisions. Be aware of the precision of *digital* instruments: they usually display more digits than warranted by their precision. The precision of reliable commercial instruments is generally indicated by the manufacturer, sometimes as an individual calibration report. Often the *maximum* error is given, which can have a (partly) systematic character and which exceeds the standard deviation.

Know where the errors are

As experimentalist you should develop a realistic feeling for the errors inherent in your experiments. Thus you should be able to focus attention on the most critical parts and balance the accuracy of the various contributing factors. Suppose you are a chemist who performs a titration by adding fluid from a syringe and weighs the syringe before and after the titration. How accurate should your (digital) weight measurement be? If the end of a titration is marked by one drop of fluid (say, 10 mg), it suffices to use a 3-decimal balance (measuring to ± 1 mg). Using a better balance wastes time and money! If you are a physicist measuring time-dependent fluorescence following a 1 ns light pulse, it suffices to analyze the emission in 100 ps intervals. Using higher resolution wastes time and money!

3.2 Error propagation

Propagation through functions

In general the required end result of a series of measurements is a *function* of one or more measured quantities. For example, if you measure the length l and width w of a rectangular plane object, both the circumference $C = 2(l + w)$ and the area $A = lw$ are (simple) functions of l and w. Assume the deviations in l and w are independent of each other with standard uncertainties Δl and Δw, respectively, what then is the standard uncertainty in C or in A? A somewhat more complicated relation is the determination of the change in standard Gibbs function ΔG^0 for an equilibrium reaction with measured equilibrium constant K:

$$\Delta G^0 = -RT \ln K, \tag{3.1}$$

where R is the gas constant and T the absolute temperature. What is the standard uncertainty in ΔG^0 given the standard uncertainty in K? And if the equilibrium constant K of a dimerization reaction $2A \rightleftharpoons A_2$ is determined by measuring concentrations $[A]$ and $[A_2]$:

$$K = \frac{[A_2]}{[A]^2}, \tag{3.2}$$

how can we determine the standard uncertainty in K given those in $[A]$ and $[A_2]$, assuming the deviations of $[A]$ and $[A_2]$ to be independent? How will the latter be modified if the deviations are *not* independent, e.g. if we measure both the total concentration $[A] + 2[A_2]$ and $[A_2]$ independently?

What we need to establish is the *propagation* of uncertainties. The clue is *differentiation*:

If the standard uncertainty in x equals σ_x, then the standard uncertainty σ_f in $f(x)$ equals

$$\sigma_f = \left| \frac{df}{dx} \right| \sigma_x. \tag{3.3}$$

Example

Consider the example of Equation (3.1) above. You have measured $K = 305 \pm 5$ at $T = 300$ K, which yields $\Delta G^0 = 14.268$ kJ/mol. The standard uncertainty $\sigma_{\Delta G}$ in ΔG now becomes $(RT/K)\sigma_K = 41$ J/mol. The result you write as $\Delta G^0 = 14.27 \pm 0.04$ kJ/mol.

Combination of independent terms

If the uncertainty in a result (e.g. the sum of two variables) is composed of uncertainties in two or more independent measured quantities, these uncertainties must be combined in an appropriate way. Simple addition of standard uncertainties cannot be correct: the deviations due to different independent

Table 3.1 *Propagation of standard uncertainties in combined quantities or functions.*

$f = x + y$ or $f = x - y$	$\sigma_f^2 = \sigma_x^2 + \sigma_y^2$
$f = xy$ or $f = x/y$	$(\sigma_f/f)^2 = (\sigma_x/x)^2 + (\sigma_y/y)^2$
$f = xy^n$ or $f = x/y^n$	$(\sigma_f/f)^2 = (\sigma_x/x)^2 + n^2(\sigma_y/y)^2$
$f = \ln x$	$\sigma_f = \sigma_x/x$
$f = e^x$	$\sigma_f = f\sigma_x$

sources can be either $+$ or $-$ and will often partly compensate each other. The correct way to "add up" uncertainties is to take the square root of the sum of the squares of the individual uncertainties. More specifically, this applies to standard deviations σ:

$$\text{If } f = x + y, \quad \text{then } \sigma_f^2 = \sigma_x^2 + \sigma_y^2, \tag{3.4}$$

i.e., *independent uncertainties add up quadratically.* Why this is so is explained in Appendix A1 on page 135. In general, when f is a function of $x, y, z, \ldots$;

$$\sigma_f^2 = \left(\frac{\partial f}{\partial x}\right)^2 \sigma_x^2 + \left(\frac{\partial f}{\partial y}\right)^2 \sigma_y^2 + \cdots \tag{3.5}$$

From (3.5) it follows immediately that for additions and subtractions the *absolute* uncertainties add up quadratically, while for multiplications and divisions the *relative* uncertainties add up quadratically. Examples of (3.5) are given in Table 3.1, valid for independent contributions.

Example 1

Consider the example of (3.2). What is the s.d. in $K = [A_2]/[A]^2$ when the deviations in $[A]$ and $[A_2]$ are independent? From the x/y^n rule in Table 3.1 it follows that

$$\left(\frac{\sigma_K}{K}\right)^2 = \left(\frac{\sigma_{[A_2]}}{[A_2]}\right)^2 + 4\left(\frac{\sigma_{[A]}}{[A]}\right)^2.$$

Suppose you have measured $[A_2] = 0.010 \pm 0.001$ mol/L and $[A] = 0.100 \pm 0.004$ mol/L. Then the *relative* s.d. of K becomes $\sqrt{0.1^2 + 4.0.04^2} = 0.13$, resulting in $K = 1.0 \pm 0.1$ L/mol.

Example 2

Consider again the example of (3.2). What is the s.d. in K if the deviations in the total concentration $[A] + 2[A_2]$ and in $[A_2]$ are independent? Rename the independent variables:

$$x = [A] + 2[A_2]; \quad y = [A_2],$$

so that

$$K = \frac{y}{(x - 2y)^2}.$$

Apply the general rule (3.5), which yields

$$\sigma_K^2 = (x - 2y)^{-6}\left(4y^2\sigma_x^2 + (x + 2y)^2\sigma_y^2\right).$$

Suppose you have measured the dimer concentration $y = 0.010 \pm 0.001$ mol/L and the total concentration of A $x = 0.120 \pm 0.005$ mol/L. Then the variance of K becomes

$$\sigma_K^2 = 400\sigma_x^2 + 19600\sigma_y^2 = 0.030.$$

So the s.d. becomes $\sqrt{0.030} = 0.17$, resulting in $K = 1.0 \pm 0.2$ L/mol.

Combination of dependent terms: covariances

When uncertainties are not independent of each other, the *covariances* between x and y play a role (see Appendix A1 for details):

$$\sigma_f^2 = \left(\frac{\partial f}{\partial x}\right)^2 \sigma_x^2 + \left(\frac{\partial f}{\partial y}\right)^2 \sigma_y^2 + 2\frac{\partial f}{\partial x}\frac{\partial f}{\partial y}\, \text{cov}\,(x, y) + \cdots \qquad (3.6)$$

See Appendix A1 on page 135 for the definition of the covariance between x and y: cov (x, y).

Systematic errors due to random deviations

When the function $f(x)$ has an appreciable curvature (second derivative) in the region of x over which the uncertainty of x spreads, a *systematic* deviation in f will occur: the expected value $E[f(x)]$ does not equal $f(E[x])$. This effect is generally in practice not very important. Appendix A2 on page 138 gives details.

Monte Carlo methods

There are cases where you can't express an explicit functional relation between a result and the factors that contribute to the result. For example, given a large number of observations of temperature, pressure, composition, etc., you predict tomorrow's weather using a forecasting model. Knowing the uncertainties of the input data, how uncertain will be the prediction? In a deterministic model (as opposed to a stochastic model) there is a functional relationship between input data and result, but it is complicated and implicit, with lots of correlations and interdependencies. The propagation of uncertainties is related to the *sensitivity* of the result for variations in each of the input data.

Here the computer comes to the rescue. When the number of input parameters is relatively small, numerical values for the derivatives, as required in (3.6), can be obtained by making a small step (preferably in both directions) for each input parameter. For a large number of input data this may not work. You may then find the uncertainty in the result by choosing many input combinations, randomly chosen from the (known) uncertainty distributions of the inputs. The computed output values will be samples of the uncertainty distribution you are looking for. Methods that use random numbers to generate results are in general called *Monte Carlo methods*.[1]

A simple example will make this clear. You are a chemist who wishes to determine the equilibrium constant of the association reaction in solution

$$A + B \rightleftharpoons AB.$$

For this purpose you dissolve 5.0 ± 0.2 mmol of substance A in 100 ± 1 mL solvent and 10.0 ± 0.2 mmol B in 100 ± 1 mL solvent; then you mix the two solutions. You determine the concentration x of AB spectroscopically (AB has an absorption band in a spectral region where neither A nor B absorbs) and find $x = 5.00 \pm 0.35$ mmol/L. The uncertainties given are all standard deviations of supposedly normal distributions. What is the value of the equilibrium constant K and what is its standard uncertainty?

The equilibrium constant is given by

$$K = \frac{[AB]}{[A][B]}, \tag{3.7}$$

---

[1] Monte Carlo methods can be applied in many fields, notably in statistics, in statistical mechanics and in mathematics to compute multidimensional definite integrals. They are used to generate samples from a given multidimensional distribution. Often a random step is followed by an acceptance criterion, allowing an efficient biased random search that concentrates on the "important" regions of the explored multidimensional space. See Hammersley and Handscomb (1964); for applications in molecular simulation see Frenkel and Smit (2002).

where [A] is the concentration of A, etc. Hence

$$K = \frac{x}{(a/(V_1 + V_2) - x)(b/(V_1 + V_2) - x)}, \qquad (3.8)$$

where a is the amount of A originally dissolved in volume V_1, b the amount of B dissolved in V_2 and x the measured concentration of AB. Of course it is quite feasible to determine K with its uncertainty from the data using the standard method based on (3.5), but it is easier to use a Monte Carlo approach. This is done by generating a large number n (e.g. $n = 1000$) of normally distributed values for each of the input variables a, b, V_1, V_2, x, using the given values for mean and standard deviation (each input variable is now an array of length n) and applying (3.8) to the arrays. The output K is an array of samples representing the probability distribution of K. Doing this we find

$$K = (5.6 \pm 0.6) \text{ L/mol.} \qquad (3.9)$$

The cumulative distribution of K is given on a probability scale in Fig. 3.1 (on a probability scale a normal distribution shows as a straight line; see page 39

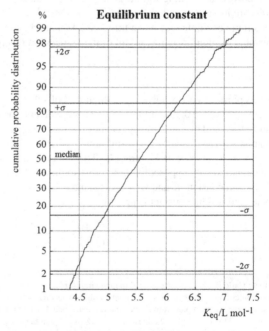

Figure 3.1 The cumulative probability distribution function of the Monte Carlo-generated result of (3.8), using 1000 samples.

for further explanation). You see that the distribution is fairly normal between $\mu \pm \sigma$, but deviates from normal beyond $\mu \pm 2\sigma$. This is due to the nonlinear relation between K and the input variables. Thus the Monte Carlo method has advantages: distortions of the resulting distribution due to nonlinearity are immediately apparent and so are systematic errors due to nonlinearity. The latter arc visible as a difference between the mean of the distribution and the value computed directly from the input values without added noise.

See **Python code** 3.1 on page 173 for the generation of Monte Carlo samples and figures for this example.

> **Summary** *This chapter distinguished between systematic and random errors, the latter leading to uncertainties in the results. Random errors add up quadratically in sums or differences (i.e., the uncertainty in the result is the square root of the sum of squares of the contributing terms). Relative random errors add up quadratically in products or quotients. Table 3.1 gives more functional relations. In general, an error in x propagates in a function f(x) through multiplication by the derivative $\partial f / \partial x$, see (3.3) and (3.5). When input errors are correlated, their covariances also play a role. When the functional relation is strongly nonlinear, random errors may cause systematic deviations. To investigate error propagation in complex cases it is advantageous to use Monte Carlo methods: generate a large number of samples of the results by randomly selecting the input parameters from appropriate probability distributions.*

Exercises

3.1 Perform the following operations and give the result with standard deviation. The standard deviations of quantities are indicated by $\pm$; they are independent of each other.
   (a) $15.000/(5.0 \pm 0.1)$
   (b) $(30.0 \pm 0.9)/(5.0 \pm 0.2)$
   (c) $\log_{10}(1000 \pm 2)$
   (d) $(20.0 \pm 0.3)\exp[-(2.00 \pm 0.01)]$

3.2 The half-life time $\tau_{1/2}$ of a first-order chemical reaction is determined at four different temperatures. The temperatures are accurate; the standard uncertainties in $\tau_{1/2}$ are indicated:

Temperature (°C)	half-life $\tau_{1/2}$ (s)
510	2000 ± 100
540	600 ± 40
570	240 ± 20
600	90 ± 10

Determine the rate constant k (what unit?) and its standard uncertainty, as well as $\ln k$ and its standard uncertainty, at every temperature. Now plot $\ln k$ with error bars versus the reciprocal absolute temperature. Also, plot k with appropriate error bars on a logarithmic scale versus the reciprocal absolute temperature. Compare the two plots.

3.3 Suppose you determine the acceleration of gravity g by measuring the oscillation period T of a pendulum with length l. The value of g follows from

$$g = 4\pi^2 l / T^2.$$

You measure $T = 2.007 \pm 0.002$ s and $l = 1.000 \pm 0.002$ m. Determine g and its standard uncertainty.

3.4 The Gibbs activation function for a chemical reaction $\Delta G^{\ddagger}$ follows from the rate constant k according to *Eyring's equation*

$$k = (k_B T / h) \exp(-\Delta G^{\ddagger} / RT).$$

Here k_B is Boltzmann's constant, h is Planck's constant and R the gas constant (see the data sheet PHYSICAL CONSTANTS on page 209, or use the Python module *physcon.py*).

(a) If the rate constant k has an uncertainty of 10%, what is the resulting uncertainty in $\Delta G^{\ddagger}$?

(b) Discuss how an uncertainty in the temperature propagates into $\Delta G^{\ddagger}$.

(c) If $\Delta G^{\ddagger} = 30$ kJ/mol and $T = 300$ K, how large is the uncertainty in $\Delta G^{\ddagger}$ as a result of an uncertainty of $5\,°C$ in the temperature?

3.5 (this exercise relates to Appendix A2 on page 138)

Generate an array with 1000 samples of the volume of spheres, of which the radii are samples of a normal distribution with mean 1.0 mm and standard deviation 0.1 mm. Compare the mean of the distribution with the volume of a sphere with radius 1.0 mm and discuss whether the latter is a biased result. Discuss the significance of the bias. Plot the cumulative volume distribution on a probability scale.

4 Probability distributions

Every measurement is in fact a random sample from a probability distribution. In order to make a judgment on the accuracy of an experimental result we must know something about the underlying probability distribution. This chapter treats the properties of probability distributions and gives details about the most common distributions. The most important distribution of all is the normal distribution, not in the least because the central limit theorem tells us that it is the limiting distribution for the sum of many random disturbances.

4.1 Introduction

Every measurement x_i of a quantity x can be considered to be a *random sample* from a *probability distribution* $p(x)$ of x. In order to be able to analyze random deviations in measured quantities we must know something about the *underlying* probability distribution, from which the measurement is supposed to be a random sample.

If x can only assume discrete values $x = k, k = 1, \ldots, n$ then $p(k)$ forms a *discrete probability distribution* and $p(k)$ (often called the *probability mass function*, pmf) indicates the probability that an arbitrary sample has the value k. If x is a continuous variable, then $p(x)$ is a continuous function of x: the *probability density function*, pdf. The meaning of $p(x)$ is: *the probability that a sample x_i occurs in the interval $(x, x + dx)$ equals $p(x)\, dx$.*

Probability density functions (or probability mass functions) are defined on a *domain* of possible values the random variable can assume. The function value itself is a non-negative real number. The integral over the domain (or the sum in the case of a discrete distribution) equals 1, i.e., the pdf (or pmf) is normalized. In general pdf's can be *multidimensional*, i.e., a function of one, two or more variables. Thus the *joint pdf* $p(x, y)$ means that the probability of finding a sample x_i in the interval $(x, x + dx)$ *and* of finding a sample y_i in the interval $(y, y + dy)$ is given by $p(x, y)\, dx\, dy$. If a pdf $p(x, y)$ is integrated over one variable, say y, the resulting pdf is called the *marginal* pdf of x; multiplied by dx it is the probability of finding x_i in the interval $(x, x + dx)$ *irrespective* of the value of y. Probabilities can also be defined under a restrictive condition, e.g. $p(x|y)$ is the *conditional* probability of finding x, *given* the value of y. The conditional probability makes sense only when x and y are somehow related to each other: if they are independent of each other, $p(x|y)$ obviously does not depend on y:

$$p(x|y) = p(x) \quad (x, y \text{ independent}). \tag{4.1}$$

The following relations hold:

$$p(x, y) = p(x)p(y|x) = p(y)p(x|y), \tag{4.2}$$

$$p(x, y) = p(x)p(y) \quad (x, y \text{ independent}), \tag{4.3}$$

where $p(x)$ and $p(y)$ are the marginal distributions:

$$p(x) = \int p(x, y)\, dy, \tag{4.4}$$

$$p(y) = \int p(x, y)\, dx. \tag{4.5}$$

The integrations are carried out over the full domains of the variables y and x.

A summary of the properties of one- and two-dimensional probability functions is given on the data sheet PROBABILITY DISTRIBUTIONS on page 211.

In this chapter we consider the properties of a few common one-dimensional probability distributions: the *binomial distribution*, the *Poisson distribution*, the *normal distribution* and a few others. The first two are discrete distributions, the latter is a continuous distribution. In the following chapter on page 53 we consider how, given a series of measured samples, we can derive the *best estimates* of properties of the underlying probability distribution. The real distribution can never be precisely determined because that would require an infinite number of samples.

We shall also change notation and denote the pdf's with $f(x)$ rather than $p(x)$. The reason is that the probability functions we consider in this chapter are based on counting the *frequencies of occurrences* of the possible outcomes given the statistical process that produces the samples. This is in contrast to more general interpretations of probabilities $p(x)$, which may include probabilities based on *beliefs* or *best estimates* considering all knowledge we have. Chapter 8 on page 111 elaborates on this point.

4.2 Properties of probability distributions

Normalization

Both continuous probability density functions $f(x)$ and discrete probability mass functions $f(k)$ are *normalized*, i.e., the sum of all probabilities (over the possible domain of sample values[1]) is equal to 1:

$$\int_{-\infty}^{\infty} f(x)\,dx = 1; \tag{4.6}$$

$$\sum_{k=1}^{n} f(k) = 1. \tag{4.7}$$

For the continuous density function $f(x)$ we have assumed that the domain of possible x-values comprises all real numbers, i.e., the interval $\langle -\infty, +\infty \rangle$, but there are also density functions with a different domain, such as $[0, 1]$ or $[0, +\infty)$. Probabilities are never negative: $f(k) \geq 0$; $f(x) \geq 0$.

Expectation, mean and variance

The *expectation* of a function $g(x)$ of x over the probability density function $f(x)$ (sometimes called the *expected value*) $E[g]$ of $g(x)$ is defined as

$$E[g] = \int_{-\infty}^{\infty} g(x)f(x)\,dx, \tag{4.8}$$

or, in the discrete case:

$$E[g] = \sum_{k=1}^{n} g(k)f(k). \tag{4.9}$$

We use the notation $E[\;]$ to indicate that E is a *functional*, i.e., a function of a function. Thus the *mean* of x, usually indicated by μ, is equal to the expectation of x itself over the density function:

$$\mu = E[x] = \int_{-\infty}^{\infty} xf(x)\,dx, \tag{4.10}$$

---

[1] The *domain* is the set of possible values of k or x; the *range* of a series of samples is the difference between the largest and the smallest value occurring in the data set. An *interval* is a set of values between a lower and an upper limit; one indicates the interval limits by [or] if the limit itself is included and by $\langle$ or $\rangle$ if the limit is not included. Normal brackets (or) may be used when the distinction is irrelevant.

or, in the discrete case:

$$\mu = E[k] = \sum_{k=1}^{n} k f(k). \tag{4.11}$$

The *variance* σ^2 of a probability distribution is the expectation of the squared deviation from the mean:

$$\sigma^2 = E[(x - \mu)^2] = \int_{-\infty}^{\infty} (x - \mu)^2 f(x)\, dx, \tag{4.12}$$

or, in the discrete case:

$$\sigma^2 = E[(k - \mu)^2] = \sum_{k=1}^{n} (k - \mu)^2 f(k). \tag{4.13}$$

The square root of σ^2 is called the *standard deviation* (s.d.) σ. Alternatively the s.d. is called the 'rms (root-mean-square) deviation'. The s.d. of the uncertainty distribution of an experimental result is called the *standard uncertainty* or *standard error* or *r.m.s. error*.

Moments and central moments

These are the most important averages over probability distributions. They are related to the first and second *moment* of the distribution. The n-th moment μ_n of a distribution is defined as

$$\mu_n = E[x^n]. \tag{4.14}$$

It is often more useful to employ the *central moments* which are defined with respect to the mean of the distribution. The n-th central moment is

$$\mu_n^c = E[(x - \mu)^n]. \tag{4.15}$$

The second central moment is the variance. The third central moment, expressed in units of σ^3, is called the *skewness* and the fourth central moment (in units σ^4) is the *kurtosis*. Since the kurtosis of a normal distribution equals 3 (see Section 4.5), the *excess* is defined as the deviation from the kurtosis of a normal distribution:[2]

$$skewness = E[(x - \mu)^3 / \sigma^3] \tag{4.16}$$

$$kurtosis = E[(x - \mu)^4 / \sigma^4] \tag{4.17}$$

$$excess = kurtosis - 3. \tag{4.18}$$

---

[2] Some books use the name *kurtosis* or *coefficient of kurtosis* for what we have defined as *excess*.

Cumulative distribution function

The cumulative distribution function (cdf) $F(x)$ gives the probability that a value x is not exceeded:

$$F(x) = \int_{-\infty}^{x} f(x')\, dx',$$ (4.19)

or, in the discrete case:

$$F(k) = \sum_{l=1}^{k} f(l).$$ (4.20)

Note that the value $f(k)$ is included in the cumulative sum $F(k)$. The function $1 - F(x)$ is called the *survival function* (sf), indicating the probability that x *is* exceeded:

$$sf(x) = 1 - F(x) = \int_{x}^{\infty} f(x')\, dx',$$ (4.21)

or, in the discrete case:

$$sf(k) = 1 - F(k) = \sum_{l=k+1}^{n} f(l).$$ (4.22)

From these definitions it is clear that

$$f(x) = \frac{dF(x)}{dx}$$ (4.23)

$$f(k) = F(k) - F(k-1).$$ (4.24)

The function F is monotonically increasing, with a value in the interval $[0, 1]$. Cumulative distribution functions F and their inverse functions F^{-1} are necessary to determine *confidence intervals* and *confidence limits*. For example, the probability that x lies between $x_1 = F^{-1}(0.25)$, i.e., $F(x_1) = 0.25$, and $x_2 = F^{-1}(0.75)$, i.e., $F(x_2) = 0.75$, is 50%. The values of x that will be exceeded with a probability of 1% is equal to $F^{-1}(0.99)$, i.e., $F(x) = 0.99$. The value $F^{-1}(0.5)$, i.e., the value of x for which $F(x) = 0.5$, is the *median* of the distribution; $F^{-1}(0.25)$ and $F^{-1}(0.75)$ are the first and third *quartiles* of the distributions. Similarly one may define *deciles* and *percentiles*. The q-th *quantile* equals x if $F(x) = q$.

Characteristic function

Every probability density function $f(x)$ has associated with it a *characteristic function* $\Phi(t)$, which is defined as

$$\Phi(t) \stackrel{\text{def}}{=} E[e^{itx}] = \int_{-\infty}^{\infty} e^{itx} f(x)\, dx.$$ (4.25)

The characteristic function is mathematically helpful to analyze probability functions. For example, its series expansion in t generates moments of the distribution. For the usual statistical data treatment you will not require the characteristic function. Interested readers, however, who are not unfamiliar with Fourier transforms, may consult Appendix A3 on page 141 for further details.

A word on nomenclature

The word *probability distribution* is often used in a general sense, meaning any kind of discrete or continuous probability function. However, sometimes the term *distribution function* is specifically meant to indicate the *cumulative distribution function* $F(x)$, in contrast to the continuous *probability density function* $f(x)$ or the discrete probability mass function $f(k)$. To avoid confusion, it is recommended that the modifier "cumulative" is included in this case. Instead of "probability distribution" you should use the term "probability density function" when the latter is meant.

Numerical values of distribution functions

Statistical tables generally give both the density functions and the cumulative functions. They can – among others – be found in Beyer (1991), in Abramowitz and Stegun (1964), and in the *Handbook of Chemistry and Physics* (CRC Handbook, each year). Values can more easily and more accurately be extracted from computer packages. The Python extension SciPy offers a package "stats" with more than 80 continuous and 12 discrete distributions; for each one can invoke the probability density function (pdf), the cumulative distribution function (cdf), the survival function (sf), the percent-point function (ppf, inverse of cdf) and the isf (inverse survival function). Random variates (rvs) and common statistical properties can be obtained from each distribution as well.

4.3 The binomial distribution

Definition and properties

Suppose you measure a *binary* quantity, i.e. a quantity that can assume either one of two values (e.g. 0 or 1, false or true) and every measurement has a probability p to be 1 (or true), then the probability $f(k; n)$ that out of n measurements exactly k have the outcome 1, equals

$$f(k; n) = \binom{n}{k} p^k (1 - p)^{n-k}. \qquad (4.26)$$

Here

$$\binom{n}{k} = \frac{n!}{k!(n-k)!} \qquad (4.27)$$

is the *binomial coefficient* "*n* over *k*" indicating the number of ways *k* objects can be chosen from a set of *n* objects. The random process to choose one possibilities out of two, with probability *p*, is called a *Bernouilli trial*. Some important properties of the binomial distribution are:

$$\text{mean: } \mu = E[k] = pn, \qquad (4.28)$$

$$\text{variance: } \sigma^2 = E[(k-\mu)^2] = p(1-p)n, \qquad (4.29)$$

$$\text{s.d.: } \sigma = \sqrt{p(1-p)n}. \qquad (4.30)$$

Appendix A4 explains why.

Variance proportional to number

The variance is proportional to the total number of observations *n* (called the *sample size*). Therefore the *relative* standard uncertainty is *inversely* proportional to the *square root* of the sample size. This is an important rule of thumb to remember: for a 100 times larger sample size the relative uncertainty becomes 10 times smaller. You can buy accuracy by doing more experiments.

Note that for small *p* the standard deviation is approximately equal to the square root of the mean number of observed events *pn*. If you have observed 100 events that only seldom occur, the s.d. in the observed number is 10, or 10%; if you have observed 1000 events, the s.d. is 32 or 3.2%. If you want to gain a factor of 10 in accuracy, your observation time must be 100 times longer.

Examples

Here are a few examples of binomial distributions. Figure 4.1 shows the probability of obtaining *k* heads in 10 coin tosses, assuming the probability of obtaining a head in each throw is 0.5. Figure 4.2 shows the probability of obtaining *k* times a "six" in 60 throws of a perfect dice. You see that the distribution tends to become symmetrical for larger numbers, even if the probability for a single event is far from the symmetrical 0.5.

Figure 4.3 relates to "extra-sensory perception" (ESP) experiments parapsychologists used to perform to investigate the possibility of telepathy.[3]

---

[3] This is a case where scientists would demand a very high significance level in order to even consider positive experimental outcomes. Previous experimenters have fallen into all the statistical pitfalls you can think of. See Gardner (1957).

Binomial 10 coin tosses

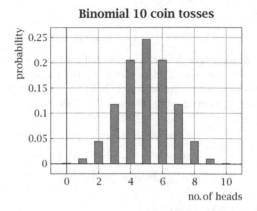

Figure 4.1 The probability of obtaining k heads in 10 coin tosses.

Binomial 60 dice throws

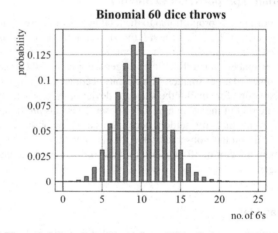

Figure 4.2 The probability of obtaining k faces "6" in 60 throws of a dice.

The "sender" sequentially selects cards from a well-mixed pack of "Zener cards," which contains an equal number of five types of card (each with a simple figure: square, circle, cross, star, wavy lines) and concentrates for a moment on the figure; the "receiver" notes which card he thinks has been drawn, without being able to see the card. One such experiment involves 25 cards. Assuming that telepathy does not exist, the probability of a correct guess is 0.2 and on average 5 cards will be guessed correctly. The probability of guessing *more than k* cards correctly is the binomial *survival function*

Table 4.1 *The binomial survival function*
$1 - F(k)$, *giving the probability that more*
than k Zener cards are guessed correctly
out of 25 trials.

$\geq (k+1)$	$>k$	survival $1 - F(k)$
12	11	0.001 540
11	10	0.005 555
10	9	0.017 332
9	8	0.046 774
8	7	0.109 123
7	6	0.219 965

Binomial 25 Zener cards

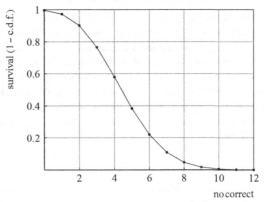

Figure 4.3 The "survival", i.e. the probability of guessing *more than k* cards correctly out of 25 trials. There are five different cards which are randomly presented.

(sf), which is 1 minus the cumulative distribution function (cdf). The exact meaning of the cdf and sf is:

$$\text{cdf}: \quad F(x): \quad \text{Prob}\{k \leq x\} = F(x); \qquad (4.31)$$

$$\text{sf}: \quad 1 - F(x): \quad \text{Prob}\{k > x\} = 1 - F(x). \qquad (4.32)$$

The survival function is given for some relevant values in Table 4.1 and in Fig. 4.3.

See **Python code** 4.1 on page 173 for codes to generate the functions and figures of this section.

From binomial to multinomial

When a random choice is made not among two possibilities, but among a number m possibilities, the statistics is that of a *multinomial* distribution. For example, an opinion poll asks which choice a voter will make among the five parties that figure in an election. Or, a certain sequence of amino acids in a protein can be classified as either $\alpha-helix$, $\beta-sheet$ or *random coil*. Or, you gather random variables in n distinct bins. The details of the multinomial distribution can be found in Appendix A4.

4.4 The Poisson distribution

You will encounter the Poisson distribution whenever you are *counting numbers*, such as a number of objects in a small volume of a homogeneous suspension (e.g. bacteria under a microscope or fish in a representative volume in a lake), or a number of photons detected in a given time interval Δt with a "single photon counter," or the number of gamma quanta counted in a given time interval originating from the radioactive decay of unstable nuclei.

If μ is the *average* number of events that can be expected, then the probability $f(k)$ of counting exactly k events is given by the Poisson distribution:

$$f(k) = \mu^k e^{-\mu}/k! \qquad (4.33)$$

The Poisson distribution is a limiting case of the binomial distribution $(p \to 0)$; for large k the Poisson distribution itself approaches a normal distribution. Details are given in Appendix A4.

The Poisson mass distribution is normalized. The mean and variance are given by:

$$E[k] = \mu, \qquad (4.34)$$

$$\sigma^2 = E[(k - \mu)^2] = \mu. \qquad (4.35)$$

The most important property of the Poisson distribution is that the standard deviation σ equals the square root of the mean μ. For example, a measurement counting 10 000 photons has a s.d. of 100, i.e. an uncertainty of 1 percent. When the number of events is sufficiently large (say, >20) then

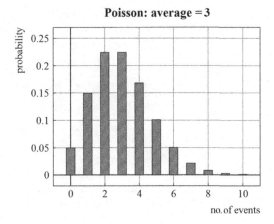

Figure 4.4 The probability that exactly k events are observed in a given time interval, when the events arrive randomly with an average of 3 per time interval.

the Poisson distribution is almost equal to the normal distribution with mean μ and s.d. $\sqrt{\mu}$.

Figure 4.4 shows the probability $f(k)$ of observing k events when the mean number $\mu = 3$. For example, a specialized hospital ward admits on the average 3 urgent patients per day; $f(k)$ is the probability that on a given day k patients arrive, assuming the patients arrive randomly. See Exercise 4.6.

4.5 The normal distribution

See data sheet NORMAL DISTRIBUTION on page 205.

The Gauss function

The pdf of the normal distribution is known mathematically as a *Gauss function*:

$$f(x) = \frac{1}{\sigma\sqrt{2\pi}} \exp\left[-\frac{(x-\mu)^2}{2\sigma^2}\right]. \tag{4.36}$$

The mean is μ, the variance is σ^2 and the s.d. is σ. The normal distribution is usually indicated by $N(\mu, \sigma)$. When we make the substitution

$$z = \frac{x-\mu}{\sigma}, \tag{4.37}$$

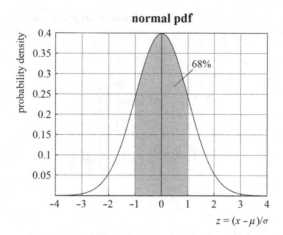

Figure 4.5 The standardized normal probability density function (pdf) $f(z)$; $z = (x - \mu)/\sigma$, with μ being the mean and σ the standard deviation of the random variable x.

the *standardized normal distribution* is obtained, indicated by $N(0, 1)$. Its density distribution is

$$f(z) = \frac{1}{\sqrt{2\pi}} \exp\left[-\frac{z^2}{2}\right]. \tag{4.38}$$

Figure 4.5 gives the standardized normal pdf. On the horizontal axis the reduced coordinate $(x - \mu)/\sigma$ is given. Thus the value 0 corresponds with $x = \mu$ and the value 1 with $x = \mu + \sigma$. The grey area gives the (integrated) probability that x lies between the values $\mu - \sigma$ and $\mu + \sigma$; this follows from the cumulative distribution function $F(z)$ and the probability equals $F(1) - F(-1) = 1 - 2F(-1) = 0.6826$ (68%).

Figure 4.6 gives the cumulative distribution function (cdf)

$$F(z) = \int_{-\infty}^{z} f(z')\, dz, \tag{4.39}$$

which expresses the probability that a sample from the normal distribution is not larger than z. The survival function (sf) $1 - F(z)$, expressing the probability that a normal variate exceeds the value z, is also given.

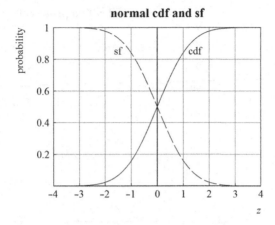

Figure 4.6 The standardized normal cumulative probability distribution function (pdf) $F(z); z = (x - \mu)/\sigma$, with μ being the mean and σ the standard deviation of the random variable x. The dashed curve is the survival function (sf) $1 - F(z)$.

Relation of cdf to error function

The function $F(z)$ can be expressed in terms of the *error function* erf (z), which is a mathematical function defined as:[4]

$$\text{erf}(x) \stackrel{\text{def}}{=} \frac{2}{\sqrt{\pi}} \int_0^x \exp(-t^2)\, dt. \qquad (4.40)$$

Its complement is the complementary error function erfc:

$$\text{erfc}(x) = 1 - \text{erf}(x). \qquad (4.41)$$

The relation is:

$$F(x) = \frac{1}{2} \text{erfc}\left(-x/\sqrt{2}\right) \quad \text{for } x < 0; \qquad (4.42)$$

$$= \frac{1}{2}\left[1 + \text{erf}(x/\sqrt{2})\right] \quad \text{for } x \geq 0. \qquad (4.43)$$

Probability scales

In order to judge whether a distribution is approximately normal, it is convenient to plot the cdf on a scale designed to produce a straight line in case of

---

[4] See, for example, Abramowitz and Stegun (1964).

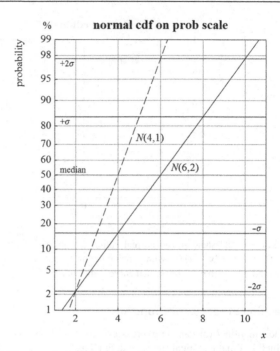

Figure 4.7 The cumulative distribution function (cdf) of a normal distribution $N(6, 2)$, i.e. $\mu = 5; \sigma = 2$ is plotted on a "probability scale" (drawn line). Dashed line: $N(4, 1)$.

normal distributions. Graph paper with appropriate divisions along the ordinate is commercially available (*probability paper*; print-yourself files can be downloaded from www.hjcb.nl/). With adequate computer software you can let the computer make the plots, rather than plotting by hand on paper. The plotting package `plotsvg` allows one to plot functions and cumulative distributions on a probability scale and such plots are often used in this book. Figure 4.7 plots two perfect normal distributions $N(6, 2)$ and $N(4, 1)$ on a probability scale: of course these are perfect straight lines. One can read the mean value and the standard deviation from such plots.

Significant deviations

Table 4.2 gives the probability that a sample x lies in a given interval and the probability that x exceeds a given value (the survival function $1 - F(z)$). You see that deviations of more than 2σ don't occur very often; deviations of more than 3σ are very rare. So if you find deviations of more than 3σ

Table 4.2 *Probability that a sample from a normal distribution occurs in the interval $(\mu - \Delta, \mu + \Delta)$ and the probability that a sample value exceeds $\mu + \Delta$ (or, equivalently, is smaller than $\mu - \Delta$), for various values of Δ.*

deviation Δ in units σ	Probability in $(\mu - \Delta, \mu + \Delta)$	Probability $> \mu + \Delta$
0.6745	50%	25%
1	68.3%	15.9%
1.5	86.6%	6.68%
2	95.45%	2.28%
2.5	98.76%	0.62%
3	99.73%	0.135%
4	99.993 66%	0.003 17%
5	99.999 943%	0.000 029%

in an experiment, you may safely conclude that is it improbable that such a deviation occurs by chance and designate the deviation as *significant*. Some researchers prefer to set the significance limit at 2.5σ or even at 2σ; what is best depends on the purpose (i.e. on the consequences of the decision taken on the basis of the measurement) and on the taste of the researcher. Of course the criterium used should always be made specific.

You should be especially careful when you consider the significance of one out of a *series* of experiments. It is not at all significant (on the contrary, it is quite likely with a probability of more than 70%) that at least one out of 100 independent measurements deviates more than 2.5σ; if you wish to maintain a significance level of e.g. 5% on the whole series of experiments, you should insist on a deviation of 3.5σ for at least 1 out of 100 results. Selecting the "significant" experiments and disregarding the "insignificant" ones, is a scientific crime. See pages 3 and 4 of the data sheet NORMAL DISTRIBUTION on page 205.

4.6 The central limit theorem

Of the various types of probability distributions the *normal distribution* is by far the most common in practice. The reason for this is that random fluctuations that are a result of the sum of many independent random components, tend to be distributed normally, independent of the type of distribution sampled by each component. This is the famous *central limit theorem*. The mean or variance of the distribution of the sum equals the sum of the means or variances of the distribution of each of the contributing components: More precisely:

Let $x_i, i = 1 \ldots n$ be a set of random variables with arbitrary probability distribution with finite mean m_i and variance σ_i^2. Then for large n the random sum variable $x = x_1 + \cdots + x_n$ tends to sample a normal distribution $N(m, \sigma)$, with

$$m = \sum_{i=1}^{n} m_i, \qquad (4.44)$$

$$\sigma^2 = \sum_{i=1}^{n} \sigma_i^2. \qquad (4.45)$$

The theorem should be used with caution: if the distribution functions of contributing components have a non-existing (infinite) variance, the central limit theorem breaks down. Heavily skewed distributions may give problems as well. Appendix A5 on page 148 gives details.

Although the central limit theorem is very important and powerful, it is not a general justification for the assumption of normality of underlying probability distributions. Relatively small deviations are often normally distributed. This is not always true for larger deviations, for example in quantities like a concentration or an intensity that can only be positive. Be aware that in such cases non-normal, skewed, distributions are likely to occur.

4.7 Other distributions

There are many other probability distributions. Some are described shortly in this section; others we shall encounter later in this book: they are important for the assessment of confidence intervals for the properties derived from data series.

Log-normal distribution

The log-normal distribution is a normal distribution of $\log x$ instead of x. It is of course only defined for $x > 0$. This distribution is especially appropriate for variables that can never be negative, such as a concentration, a length, a volume, a time interval, etc.

The standard form for the density distribution function, as available in Python through the SciPy function `stats.lognorm.pdf`, is

$$f_{st}(x, s) = \frac{1}{sx\sqrt{2\pi}} \exp\left[-\frac{1}{2}\left(\frac{\ln x}{s}\right)^2 \right], \qquad (4.46)$$

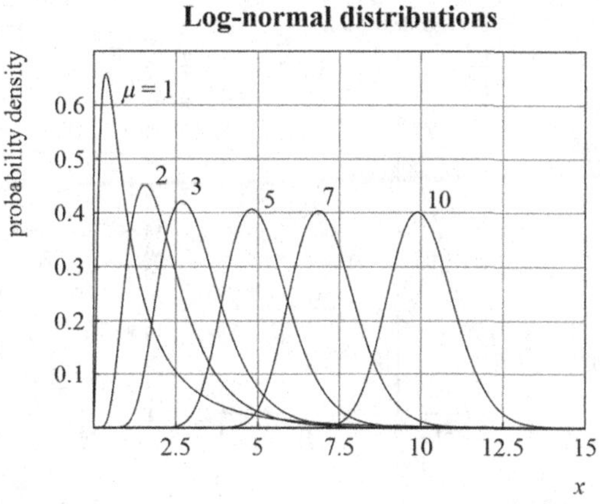

Log-normal distributions

Figure 4.8 The probability density function (pdf) of log-normal distributions $f(x; \mu, \sigma)$, see (4.47), for various values of μ. The value of $\sigma = 1$ for all curves.

but a more convenient form is

$$f(x; \mu, \sigma) = \frac{1}{\mu} f_{st}\left(\frac{x}{\mu}, \frac{s}{\mu}\right). \qquad (4.47)$$

In this form $f(x; \mu, \sigma)$ approaches the normal density function $N(\mu, \sigma)$ when μ/σ becomes larger. Figure 4.8 shows examples of the log-normal pdf with various μ, but all with the same $\sigma = 1$. For $\mu = 10\sigma$ the shape of the curve is virtually indistinguishable from the normal pdf.

The Lorentz distribution: undefined variance

A somewhat unusual distribution, but one with special interest, is the *Lorentz distribution*, also known by the name *Cauchy distribution*:

$$f(x; \mu, w) = \frac{1}{\pi w}\left[1 + \left(\frac{x - \mu}{w}\right)^2\right]^{-1}, \qquad (4.48)$$

where μ is the mean and w is a width parameter. At $x = \mu \pm w$ the function is at half its maximum height. A measure for the width is the FWHH (full width at half height), equal to $2w$. This distribution may arise from spectroscopic experiments: the frequency distribution of emitted quanta from a sharp lifetime-limited excited state has a Lorentzian shape. The Lorentzian shape also arises in another context: Student's t-distribution for one degree of freedom (see data sheet STUDENT'S T-DISTRIBUTION on page 213).

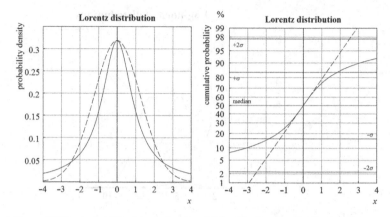

Figure 4.9 The probability density function (pdf) of the Lorentz distribution (drawn lines) $f(x; 0, 1)$, see (4.48), compared with a normal distribution (broken lines) at the same maximum height of the pdf, i.e. the same slope of the cdf at the median $(\sigma = \sqrt{\pi/2})$. Left: pdf, right: cdf on a probability scale.

The cumulative distribution is

$$F(x) = \frac{1}{2} + \frac{1}{\pi} \arctan \frac{x}{w}. \tag{4.49}$$

The problem with this distribution is that it has an infinite variance. Thus it makes no sense to estimate its variance from an actual data set. For distributions like this, including other distributions with wide tails, one should use *robust* methods (see Section 5.7 on page 63) to assess the accuracy of the mean of a series of measured samples.

Figure 4.9 depicts the Lorentz distribution, together with a normal distribution fitted with the same maximum of the pdf.

Lifetime and exponential distributions

Special types of distribution arise from considering lifetime distributions. For example, consider a large batch of incandescent lamps, all new from the factory. At time $t = 0$ you switch them all on and note the moment each lamp fails. The fraction of lamps that fails between t and $t + \Delta t$ (or equivalently, the fraction that has a lifetime between t and $t + \Delta t$) is $f(t) \Delta t$ (for small Δt), where $f(t)$ is the probability density function for the lifetime distribution. The cumulative distribution function $F(t) = \int_0^t f(t') \, dt'$ is the fraction that failed up to time t, and the survival function $1 - F(t)$ is the fraction that survives at time t (i.e., the fraction that has not (yet) failed). Another example is the lifetime distribution of individuals in a population. Consider a large number of individuals and set for each $t = 0$ at birth; $f(t) \Delta t$ is the fraction with life span between t and $t + \Delta t$; $F(t)$ is the fraction with life span $\leq t$; $1 - F(t)$ is the fraction that survives at time t. An example from the molecular sciences is the time dependence of fluorescent intensity (emitted radiation quanta) after a fluorescent molecule has been excited by a short laser pulse at $t = 0$; $f(t)$ is the normalized time-dependent intensity.

The hazard function

The lifetime probability density or its cumulative distribution function describes the lifetime statistics, but does not describe the basic cause of death or failure. More basic is the *hazard function* (also called the *failure rate function*) $h(t)$. The hazard function is the probability density that a member of the population with age t will fail (die, drop out). In other words, the probability that a member fails in a small time interval Δt around t equals $h(t) \Delta t$. Because only a fraction $1 - F(t)$ is present in the population at time t, the following relation holds:

$$h(t) = \frac{f(t)}{1 - F(t)}. \qquad (4.50)$$

From this relation, and using the fact that f is the derivative of F, we can solve for the lifetime density function:

$$f(t) = h(t) \exp\left[-\int_0^t h(t')\, dt' \right]. \qquad (4.51)$$

The exponential distribution

Several distribution functions result from various choices of $h(t)$. By far the simplest choice, which describes quite common phenomena in physics or chemistry such as radioactive decay and first-order chemical reactions, is

$$h(t) = k \quad \text{(constant)}, \qquad (4.52)$$

called the *rate constant*. Its meaning is the relative fraction of the population members (e.g. number of radioactive nuclei n, concentration of reactant c, etc.) that disappear per unit of time

$$\frac{dn}{dt} = -kn, \qquad (4.53)$$

$$\frac{dc}{dt} = -kc. \qquad (4.54)$$

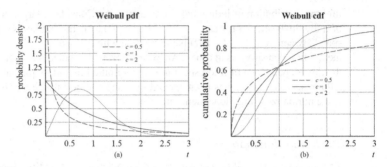

Figure 4.10 The distribution functions (pdf and cdf) of three Weibull distributions with $c = 0.5$, 1, 2. For $c = 1$ the exponential distribution is obtained.

It now follows from (4.51) that

$$f(t) = ke^{-kt} \tag{4.55}$$

and

$$F(t) = 1 - e^{-kt}. \tag{4.56}$$

This is an *exponential distribution*. The exponential distribution ($c = 1$) is depicted in Fig. 4.10

Population statistics

For the purpose of population statistics, e.g. for human population dynamics or for failure analysis, various general forms for the hazard functions have been proposed, leading to more general probability density functions for populations. The *Weibull* distribution[5] is a generalized form of the exponential distribution: the hazard function has the form

$$h(t) = ct^{c-1}. \tag{4.57}$$

Here c sets the time dependence of the failure rate; $c = 1$ recovers the exponential pdf, $c < 1$ means a higher initial rate (like a high infant mortality) and $c > 1$ means a higher failure rate at older age. The corresponding pdf is

$$f(t) = ct^{c-1} \exp[-t^c] \tag{4.58}$$

---

[5] A valuable source for information on distributions is the on-line NIST/SEMATECH e-Handbook of Statistical Methods on www.itl.nist.gov/div898/handbook.

and the cumulative distribution (cdf) is

$$F(t) = 1 - \exp[-t^c]. \tag{4.59}$$

Additional location (translating t) and scale (scaling t) parameters may be included. Figure 4.10 gives a few examples of Weibull distributions, including the exponential distribution.

See for the generation of Weibull distribution functions **Python code** 4.2 on page 174.

Chi-squared distribution

This is the distribution of the sum χ^2 of the squares of a number of normally distributed variables. The χ^2-distribution is used to obtain confidence intervals for predicted values when the s.d. of the data is known. See Section 7.4 on page 95 and the data sheet CHI-SQUARED DISTRIBUTION on page 199.

Student's t-distribution

This is the distribution of the ratio of a normally distributed variable and a χ^2-distributed variable. The t-distribution is used to assess confidence intervals for the mean, given a series of normally distributed data, when the s.d. of the distribution is not known beforehand. See Section 5.4 on page 59, the second example of Section 8.4 on page 115 and the data sheet STUDENT'S T-DISTRIBUTION on page 213.

F-distribution

This is the distribution of the ratio of two χ^2-distributed variables. The *F-ratio* is the ratio between two mean sum of squares (i.e., the sum of square deviations of a set of samples with respect to their average or with respect to a predicted value, divided by the number of degrees of freedom ν). The F-distribution (named by Snedecor) is the cumulative distribution function of the F-ratio F_{ν_1,ν_2} for the case that both sets of samples come from distributions with the same variance. It is usual to take the ratio as the largest value divided by the smallest value; if F_{ν_1,ν_2} exceeds the 99 percent level, the probability that both sets of samples come from the same distribution is less than 1 percent. The equation for F_{ν_1,ν_2} and a short table are given in data sheet F-DISTRIBUTION on page 201.

The F-distribution is useful in linear regression (see Chapter 7) in order to assess the relevance of the model that is fitted to the data. It compares the variance in the data as explained by the model with the remaining variance of the data with respect to the model; the cumulative probability of the F-ratio then indicates whether the model contributes significantly to the explanation of the data variance.

The use in regression is a special case of the general "analysis of variance" (ANOVA), which is widely used in the assessments of the influence of external factors on a normally distributed variable. Such assessments belong to the statistical domain of *experimental* or *factorial design*: the analysis of the influence of a designed external factor. As this book concentrates on the processing of data to estimate probability distributions of parameters, the statistical treatment of experimental design falls outside of its scope.[6] However, in order to give some insight into the use of F-distributions, a simple one-way ANOVA example is given below.

A group of patients, randomly selected from a homogeneous population, is treated with a drug, while another group, randomly selected from the same population, is treated with a placebo. The groups are compared by measuring an objective test value and a statistical test is performed to assess the probability that the drug treatment has been effective. The assessment is phrased in terms of the probability that the *null hypothesis* H_0 = "the drug has no influence" is true. One computes two types of mean squared averaged deviations: first of the averages of each group with respect to the global average ("between-groups variance" or mean of the "regression sum of squares" SSR) and second of the values within each group with respect to the average of that group, added over all groups ("within-group variance" or mean of "error sum of squares" SSE). Each sum of squares is divided by the number of degrees of freedom ν, i.e., the number of samples minus the number of adjustable parameters. For the "between-groups variance" $\nu_1 = k - 1$ when there are k groups; for the "within-group variance" $\nu = n - k$. In this example there are two groups: $k = 2$, one control group with n_1 observations and average μ_1, and one treated group with n_2 observations and average μ_2. The overall average of $n = n_1 + n_2$ observations y_i is μ. The *F-ratio* is

$$F_{1,n-2} = \frac{\text{SSR}/1}{\text{SSE}/(n-2)}, \tag{4.60}$$

where

$$\text{SSR} = n_1(\mu_1 - \mu)^2 + n_2(\mu_2 - \mu)^2; \tag{4.61}$$

$$\text{SSE} = \sum_{i=1}^{n_1} (y_i - \mu_1)^2 + \sum_{i=n_1+1}^{n} (y_i - \mu_2)^2. \tag{4.62}$$

---

[6] There are many books covering factorial design, e.g. Walpole *et al.* (2007).

The F-test – in fact, the value $1 - F(F_{1,n-2})$ – now tells you what the probability is that *at least this ratio* would be found if your null-hypothesis were true. If this value is small (say, less than 0.01), you may conclude that the treatment has a significant effect.

Example

Imagine that you are a physician and you want to test a new drug for treating patients with high blood pressure. You select a group of ten patients with high blood pressure who do not (yet) receive treatment and who all have agreed to participate in your trial. You design a standard way to determine the blood pressure (e.g. the average of systolic pressures at 9 am on five consecutive days) and define the test value e.g. as the blood pressure after two weeks of treatment minus the value before treatment. Then you select five patients randomly to form the "treatment group"; the remaining five patients form the control group. The treatment group receives the drug treatment and the control group receives an indistinguishable placebo. You will accept the treatment as effective when the null hypothesis is rejected at a 95 percent confidence level.[7] The outcome of the experiment (the test values in mm Hg) is as follows:

treatment group: $-21, -2, -15, +3, -22$
control group: $-8, +2, +10, -1, -4$

The treatment group has an average of -11.5 and the control group has an average of -0.2. This looks like a positive result, but if you evaluate the appropriate sums you obtain:

$$SSR = 314; \ SSE = 698; \ F_{1,8} = [314/1]/[698/8] = 3.59; \ F(3.59) = 0.91.$$

This means that there is a 9 percent probability that the null hypothesis ("the treatment has no effect") is true and a 91 percent probability that the alternative hypothesis ("the treatment is effective") is true. So, although the result suggests that the treatment is effective, you cannot come to that conclusion

---

[7] It is important that you define all experimental details *and* the statistical methods to be used *before* you do the experiment without changing your methods during or after the experiment. The selection of patients and the performance of the measurements must be completely unbiased. In a serious experiment neither the patient nor the physician who performs the measurements is allowed to know which of the patients receive the treatment (a *double-blind* experiment). A serious experiment should involve a much larger group and include safeguards when intolerable side effects occur or when the treatment appears to be so effective that it would be ethically unacceptable to deprive the control group from the benefits of treatment. A serious hospital or research organization will set rules for such experiments on humans and establish an ethical approval committee. A serious journal will evaluate the quality of the experiment before publishing the results.

when you adhere to your preset 95 percent confidence level! What you have to do, of course, is to repeat your experiment with a (much) larger number of patients.

Summary *You now have distinguished probability density distributions, cumulative probability distributions and survival functions. You know what the expectation of a function over a given distribution is and you know how the mean, the variance, the standard deviation, the skewness and the kurtosis of a distribution are defined. The binomial distribution is the simplest discrete distribution; it is suitable to describe the random picking of one out of two unequal possibilities. Random picking of one out of several possibilities is described by the multinomial distribution. Random picking of an event on a continuous scale, as the time at which an impulse or photon is observed, leads to the Poisson distribution. In the limit of many events the latter leads to a continuous Gaussian or normal distribution. The normal distribution is quite common; it emerges when a deviation is composed of many independent random contributions, irrespective of the individual distributions for each of the contributions (the central limit theorem). Some other distributions play a role in special applications; lifetime distributions are an important subclass. Distributions that have infinite variance, such as the Lorentz distribution, may cause trouble because common rules do not apply. The chi-square, Student's t- and Snedecor's F-distributions play a role in the evaluation of data series.*

Exercises

4.1 In a lottery 5 percent of the tickets will produce a prize. If you buy ten tickets, what is the probability that you obtain no prize, 1 prize, 2 prizes, ...? Assume that there are so many tickets and prizes that the probability of obtaining a prize does not depend on the number of prizes you already have (this is called: a lottery with replacement).

4.2 When it is known that one measurement x has a probability of exceeding a given value x_m of 1 percent what then is the probability that *at least* one measurement in a series of 20 independent measurements will exceed x_m?

4.3 There will be elections where voters can elect one of two presidential candidates. You want to perform an opinion poll and predict the outcome with a standard uncertainty of 1 percent. You expect roughly equal votes for either candidate. Assume that you are able to obtain the opinion of an unbiased random selection of voters, how many people do you have to select (what should be your sample size)?

4.4 You observe n independent events, each of which can have an outcome of 0 or 1. You count k_0 zeros and k_1 ones ($k_0 + k_1 = n$).

(a) What is your best estimate of the probability that a one appears?

(b) Give an estimate for the standard uncertainty in k_0.

(c) What is the standard uncertainty in k_1?

(d) You are finally interested in the ratio $r = k_1/k_0$. What is the standard uncertainty in r?

4.5 Show that the Poisson function 4.33 is normalized.

4.6 (a) With the hospital example of Fig. 4.4: assume each patient occupies a bed for one day and the ward has seven beds. When more than seven patients arrive, the excess is transported to another hospital. How many beds are occupied on average?

(b) How many patients per day are transported on average?

(c) If an unoccupied bed costs $ 300 per day and transporting one patient costs $ 1500, financially optimize the number of beds. How many patients per day are transported in the optimized case?

4.7 (a) A photosensitive device produces one electrical impulse for every absorbed photon, but also produces impulses when there is no light (the "dark current"). The number of impulses counted in 1 s is 100 without radiation and 900 with radiation. How large is the relative standard uncertainty in the measured radiation intensity?

(b) How large will be the relative standard uncertainty in the measured radiation intensity when the measurement (with and without radiation) is repeated 100 times?

4.8 What is the probability that a sample from a normally distributed quantity lies in the interval $[\mu - 0.1\sigma, \mu + 0.1\sigma]$?

4.9 (See the data sheet NORMAL DISTRIBUTION on page 205)
Using the approximation for large x mentioned on page 2 of the data sheet NORMAL DISTRIBUTION, determine the probability that the value $x = 6\sigma$ is exceeded. Is this approximation valid for this case?

4.10 The central limit theorem has a useful application: By adding 12 random numbers r, which are uniformly distributed over the interval $[0, 1)$, and subtracting 6 from the sum, you obtain in a good approximation a sample from a normal distribution with $\mu = 0$ and $\sigma = 1$:

$$x = \sum_{i=1}^{12} r_i - 6$$

(a) Show that $\langle x^2 \rangle = 1$.

(b) Generate a list of 100 normally distributed numbers by this method.

(c) Plot the cdf of this list on a "probability" scale.

4.11 Compute the mean and variance of the exponential distribution.

4.12 Refer to the example on page 49. In a similar trial the following results were obtained:

treatment group: $-6, 2, -8, -7, -12$
control group: $5, -1, 3, -4, 0$

Compute the F-ratio and the corresponding cumulative probability using the F-distribution. What conclusions would you derive from this F-test?

5 Processing of experimental data

This chapter is about the processing of data in its simplest form: given a number of similar observations $x_i = \mu + \epsilon_i$ of an unknown quantity μ, yielding values that only differ in their random fluctuations ϵ_i, how can you make the best estimate $\hat{\mu}$ of the true μ? And how can you best estimate the *accuracy* of $\hat{\mu}$, i.e., how large do you expect the deviation of $\hat{\mu}$ from the true μ to be? Each observation is a sample from an underlying distribution; how can you characterize that distribution? If you have reasons to assume that the underlying distribution is normal, how do you estimate its mean and variance and how do you assess the relative accuracy of those parameters? And how do you proceed if you don't wish to make any assumptions about the underlying distribution?

Suppose you have a number of similar observations x_i, differing only in deviations of random character. *Assume* for the time being that the probability distribution of those random deviations is a normal distribution, characterized by a mean μ and a standard deviation σ or variance σ^2. Although you don't know the real distribution function because your data set is limited (and these data are only samples from the distribution), it is possible to make *estimates* of μ and σ. Such estimates are often indicated by a *hat* over the symbol: $\hat{\mu}, \hat{\sigma}$. What you really want to know is the best estimate for the true value (e.g. the mean) and the uncertainty in that estimate. In practice you can use the estimated variance of the distribution to derive the uncertainty in the mean.

In this chapter we shall first look at the distribution function of the data (Section 5.1) and then indicate how the properties of the data (Section 5.2) lead to estimation of the properties of the distribution function (Section 5.3). The uncertainties in the estimates are discussed for the mean in Section 5.4 and for the variance in Section 5.5. Section 5.6 considers the case that individual data have different statistical weights. Finally, Section 5.7 treats some methods that are robust against dependency on the exact shape of the underlying distribution.

53

5.1 The distribution function of a data series

In order to get an impression of the distribution of the data it is useful to plot the data in a *histogram*. This is done by first sorting the data in increasing order and subsequently grouping the data in predetermined intervals. A plot of the number of observations in each interval, e.g. as bars, versus the central values of the intervals is called a histogram.

Be careful with computer programs that generate fancy histograms. For example, if you display a perspective view using three-dimensional bars, the point of view may be chosen such that certain bars appear relatively larger than they are. Horizontal lines may appear as having positive or negative slopes, and the reader may be misled by the graph. The same happens when icons are used instead of lines or bars, e.g. an oil barrel to indicate the volume of oil production: a barrel that is twice as large gives the impression of an increase much larger than a factor of two. It is naive to use fancy displays for esthetical reasons if they produce misleading results; it is a scientific crime to purposely construct misleading displays.[1]

Let us use the example "Thirty Observations" given in Chapter 2 on page 6. The data, already sorted, are given in Table 2.1 on page 6 and a histogram is shown in Fig. 2.3.

A histogram is an approximation to the probability density function that is sampled by the data. When the number of observations is limited, as in the present example of thirty observations, a histogram is quite noisy and it is difficult to judge the probability density function from a fit to the histogram. It is then much better to display the *cumulative distribution* of the data. This is quite similar to the cumulative distribution function $F(x)$ of a continuous probability distribution, as described in the previous chapter (see page 31), except that the data are now discrete. For a set of n values $x_1, \ldots, x_n$, the cumulative distribution $F_n(x)$ is defined as

$$F_n(x) = \frac{1}{n} \sum_{i=1}^{n} I(x_i \leq x), \qquad (5.1)$$

where $I(condition)$ is the *indicator function*, defined as equal to 1 when *condition* is true and equal to 0 otherwise. Thus $F_n(x)$ is equal to the fraction of all samples x_i for which $x_i \leq x$. So, between x_{i-1} and x_i the function is equal to $(i-1)/n$, but it jumps to i/n for $x = x_i$. See Fig. 5.1.

For a series of measurements with equal weight, the cumulative distribution is constructed by plotting the sequence number in an *ordered* series of data $x_1 \leq x_2 \leq \ldots \leq x_n$ versus the value of x.

The definition (5.1) assumes that all data points have equal statistical weights. This is often not the case. For example, the data may have been

---

[1] There is nothing new in misleading your readers. For examples, see Huff (1973).

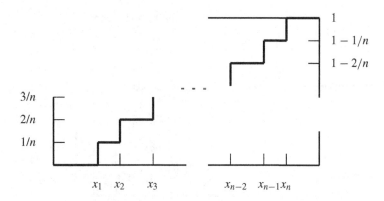

Figure 5.1 Detailed aspects of the cumulative distribution function of a set of discrete data $x_1, \ldots, x_n$.

gathered in bins before analysis (resulting in a histogram) and the individual original data are not available anymore. In that case we have, instead of n points each with statistical weight $1/n$, n bins each with a given statistical weight w_i. The latter is the number of observations within the i-th bin, preferably relative to the total number of observations, so that the total weight equals 1.

Figure 5.2 is an example of such a histogram. The data are the distribution of the height of men and women in the Netherlands in the age group 20–29, averaged over the years 1998, 1999 and 2000. The data are available from official statistical sources[2] but only in the form of percentages in bins of 5 cm width. The bin with midpoint 180 cm accumulates the rounded heights 178–182, i.e., all heights between 177.5 and 182.5 cm. The corresponding bar in the histogram should be centered at the midpoint value.

The definition of the cumulative distribution of the data is now slightly different from (5.1), as each point must be scaled according to its weight w_i:

$$F_n(x) = \frac{\sum_{i=1}^{n} w_i I(x_i \le x)}{\sum_{i=1}^{n} w_i}. \tag{5.2}$$

The data should be plotted with the "jumps" located at the midpoints of the bins. The left panel of Fig. 5.3 plots the population–height data of Fig. 5.2 as a cumulative distribution. The staircase curve, of course, is an approximation to the exact cumulative length distribution. The dots in this figure denote the points at which this approximate curve coincides with the exact

[2] http://statline.cbs.nl/StatWeb/publications.

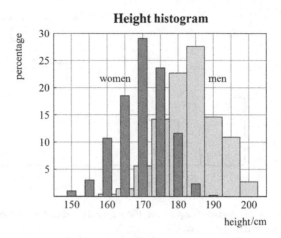

Figure 5.2 Histograms of the height distribution of men (light gray) and women (dark gray) in the age group 20–29 in the Netherlands, averaged over the years 1998, 1999 and 2000. The data have been gathered in bins of 5 cm width.

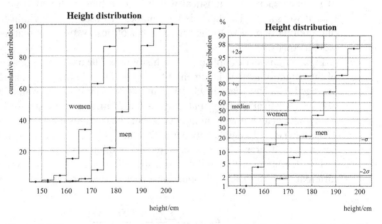

Figure 5.3 The cumulative probability distribution of the data of Fig. 5.2. Left: linear scale; right: probability scale. The dots indicate the values where the cumulative function is exact.

cumulative distribution. These points are located at the boundaries between bins. Thus, if you want to fit a theoretical distribution function to the experimental data, the theoretical curve should match these points as closely as possible.

The right panel of Fig. 5.3 plots the same data on a probability scale (see page 39). A normal distribution should give a straight line. It is obvious from this plot that the distribution of the data is very nearly normal.

5.2 The average and the mean squared deviation of a data series

In this book we denote *averages* over a data series with $\langle \ldots \rangle$ (e.g. $\langle x \rangle$).[3] In order to estimate the properties of the probability distribution from which the data are samples, the following averages are needed:

(i) The *average* $\langle x \rangle$ of a series of equivalent (i.e., equally probable) independent samples $x_i, i = 1, \ldots, n$ is given by

$$\langle x \rangle = \frac{1}{n} \sum_{i=1}^{n} x_i. \tag{5.3}$$

See Section 5.6 for the handling of data series with unequal statistical weights.

(ii) The *mean squared deviation* (msd) from the average is defined as

$$\langle (\Delta x)^2 \rangle = \frac{1}{n} \sum_{i=1}^{n} (\Delta x_i)^2, \tag{5.4}$$

where Δx_i is the deviation of the average:

$$\Delta x_i = x_i - \langle x \rangle. \tag{5.5}$$

The root of the msd, which is naturally called the *root-mean-squared deviation* (rms deviation or rmsd), is a measure for the spread of the data around the average.

In order to determine the msd, you must pass through the data twice: first to determine $\langle x \rangle$ and subsequently to determine $\langle (\Delta x)^2 \rangle$. This can be avoided by using the following identity (see Exercise 5.2):

$$\langle (\Delta x)^2 \rangle = \langle x^2 \rangle - \langle x \rangle^2, \tag{5.6}$$

where

$$\langle x^2 \rangle = \frac{1}{n} \sum_{i=1}^{n} x_i^2. \tag{5.7}$$

---

[3] Often averages are denoted with a bar over the variable, e.g. $\bar{x}$; this symbol we shall reserve for *averages over time*. The *expectation* (see page 29) is also an average, e.g. over a probability density function; this kind of average is usually named the *mean*. In the literature the term *mean* is often also employed for averages over data series.

Note: If the x_i's are large numbers with a relatively small spread, (5.6) could give inaccurate results by truncation errors, especially on a computer with single-precision arithmetic. Therefore the general use of (5.6) is not recommended. The remedy is to subtract from all x values a constant which is close to $\langle x \rangle$, e.g. the first value of the series. The computed average must of course be corrected for this shift.

5.3 Estimates for mean and variance

The averages $\langle x \rangle$ and $\langle (\Delta x)^2 \rangle$ are simple properties of the data set. We wish to use those to *estimate* the mean and variance (and hence also the standard deviation) of the underlying probability distribution of which the data are supposed to be *independent* random samples.

For the mean μ the answer is simple: the best estimate $\hat{\mu}$ for the mean of the underlying distribution is the average of the data themselves:

$$\hat{\mu} = \langle x \rangle. \tag{5.8}$$

It is easy to show that this choice for $\hat{\mu}$ minimizes the total squared deviation from $\hat{\mu}$:

$$\sum_{i=1}^{n} (x_i - \hat{\mu})^2 \text{minimal}. \tag{5.9}$$

For the variance the choice is less straightforward. The best estimate $\hat{\sigma}^2$ for the variance of the underlying distribution is slightly larger than the mean squared deviation of the average of the data:

$$\hat{\sigma}^2 = \frac{n}{n-1} \langle (\Delta x)^2 \rangle = \frac{1}{n-1} \sum_{i=1}^{n} (x_i - \langle x \rangle)^2. \tag{5.10}$$

The best estimate for the standard deviation (s.d.) of the underlying distribution is the square root of $\hat{\sigma}^2$:

$$\hat{\sigma} = \sqrt{\hat{\sigma}^2}. \tag{5.11}$$

The reason that the factor $n/(n-1)$ figures in (5.10) is that $\langle x \rangle$ is not exactly equal to the mean of the distribution, but is itself correlated with the data. One could loosely say that one data point has been "used" to compute the average, so that only $n-1$ points provide new data to compute the variance. For a derivation of this term see Appendix A6 on page 151. The equation for $\hat{\sigma}^2$ is only valid when the data are independent samples (which we assumed to be the case). When the data are correlated, $\hat{\sigma}^2$ is even larger. As you can see, the factor $n/(n-1)$ is not very important when n is large.[4]

5.4 Accuracy of mean and Student's t-distribution

The accuracy of the mean does not equal σ, but it does follow from the value of σ. The more data points are available, the more accurately the average of the measured values will represent the true mean of the underlying distribution. The average $\langle x \rangle$ is itself also a sample from a probability distribution; we could recover that distribution if we could repeat the whole series of measurements a larger number of times. When many series of n independent measurements had been performed, the variance of the average would be given by

$$\sigma_{\langle x \rangle}^2 = \sigma^2/n. \tag{5.12}$$

See Appendix A7 for the derivation of this equation. Thus the *estimate* $\hat{\sigma}_{\langle x \rangle}$ of the standard deviation of the average $\langle x \rangle$ (also called the *standard error* or *rms error* of $\langle x \rangle$) is:

$$\hat{\sigma}_{\langle x \rangle} = \frac{\hat{\sigma}}{\sqrt{n}} = \sqrt{\frac{\langle (\Delta x)^2 \rangle}{n-1}}. \tag{5.13}$$

Also this equation is only valid when the statistical deviations in all measured values are independent. If they are not independent, the individual fluctuations will not add quadratically and the standard error will become *larger*. It is as if the number of independent points is less than n. For the common case that dependencies in a series of measurements result from correlation between successive points it is possible to define a *correlation length* n_c. The equations then remain valid, but the number of data points n must sometimes be replaced by the *effective number* n/n_c. For example, in (5.10) $n/(n-1)$ must be replaced by $n/(n-n_c)$, making the estimate of the variation somewhat larger. But the standard inaccuracy in the sample mean becomes $\sqrt{n_c}$ times larger, as n in (5.12) must be replaced by n/n_c. See Appendices A6 on page 151 and A7 on page 154 for more details.

When the measurements are samples from a normal distribution, one might well expect that the quantity

$$t = \frac{\langle x \rangle - \mu}{\hat{\sigma}/\sqrt{n}} \tag{5.14}$$

---

[4] Note that calculators with statistical functions often let you choose between a σ based on n and a σ based on $n-1$. The former gives the rmsd of the data set and the latter gives the best estimate of the standard deviation of the underlying probability distribution.

will be a sample from the standard normal distribution $N(0, 1)$. This, however, is not the case because $\hat{\sigma}$ is not exactly equal to the true σ of the distribution; there is also a spread in $\hat{\sigma}$ itself. If this is taken into account, then one finds that t is a sample from a distribution called the *Student's t-distribution*.[5] For details see the data sheet STUDENT'S T-DISTRIBUTION on page 213. For a derivation in a Bayesian context see the second example in Section 8.4 on page 115.

In the limit of large numbers of data points the t-distribution equals a normal distribution, but for small numbers the t-distribution is broader. The t-distribution has as parameter the number of *degrees of freedom* $v = n - 1$, one less than the number of (independent) data points. One data point has already been "used" to determine the average, just as in the case of the estimation of σ, see (5.10). It is clear that it is only possible to say anything about the accuracy of the mean when at least two data points are available.

When the t-distribution is used, one can best give a *confidence interval*, e.g. the lower and upper limits between which the true mean is expected to lie with a probability of 50% (or 80%, 90%, 95%, 99%, ..., your choice!).

5.5 Accuracy of variance

Finally we give an indication for the accuracy of $\hat{\sigma}$: if the measurements are independent and the deviations are random samples from a normal distribution, then the *relative* standard inaccuracy of $\hat{\sigma}$ equals $1/\sqrt{2(n-1)}$. Appendix A7 on page 154 gives more details. The same applies to the relative standard inaccuracy in the computed standard error of the mean. For example, if you find for the estimated mean of a series of 10 measurements and its estimated inaccuracy 5.367 ± 0.253 than you should report this as 5.4 ± 0.3 because the relative inaccuracy of the number 0.253 equals $1/\sqrt{18}$ or 24% ($=0.06$) (insufficiently accurate for two significant digits). Had these numbers been the result of 100 independent measurements, then the proper report would have been 5.37 ± 0.25. Table 5.1 gives the relative s.d. as a percentage of $\hat{\sigma}$ for various numbers n of independent data points. The same relative inaccuracy also applies to the s.d. of the mean, as calculated by (5.13).

While the accuracy of the standard deviation is usually not very large, the estimated *skewness* or *excess* is often hardly significant. For near-Gaussian distributions these estimates with their s.d. are

---

[5] See Gosset (1908). "Student" was the pseudonym of the English statistician W. S. Gosset (b. 1876).

Table 5.1 *Relative inaccuracy (s.d.) of the estimated*
standard deviation $\hat{\sigma}$ of a distribution based on a series
of n independent samples.

n	s.d.($\hat{\sigma}$) %	n	s.d.($\hat{\sigma}$) %	n	s.d.($\hat{\sigma}$) %
2	70	10	24	50	10.1
3	50	15	19	60	9.2
4	41	20	16	70	8.5
5	35	25	14	80	8.0
6	32	30	13	90	7.5
7	29	35	12	100	7.1
8	27	40	11	150	5.8
9	25	45	11	200	5.0

$$skewness = \frac{1}{n}\sum_{i=1}^{n}\left(\frac{x_i}{\hat{\sigma}}\right)^3 \pm \sqrt{\frac{15}{n}}, \tag{5.15}$$

$$excess = \frac{1}{n}\sum_{i=1}^{n}\left(\frac{x_i}{\hat{\sigma}}\right)^4 - 3 \pm \sqrt{\frac{96}{n}}. \tag{5.16}$$

5.6 Handling data with unequal weights

Until this point we have assumed that all data points have the same *statistical weight*, i.e., that they are all samples from the same probability distribution. But it is quite common that one measurement is more accurate than another; in such cases the more accurate measurement must get a larger weight in the statistical analysis (e.g. in the determination of the mean) than a less accurate measurement. This may happen when the same quantity is determined in different ways, yielding several values with their individual uncertainty estimates, and the best estimate for the mean is required. Unequal weights must also be given to histogram data that result from adding observations in bins: it is obvious that each bin (central) value x_i must be multiplied by the number of observations in that bin n_i in order to obtain the proper mean over all observations:

$$\langle x \rangle = \frac{\sum_i n_i x_i}{\sum_i n_i}. \tag{5.17}$$

In general, the best estimate $\hat{\mu}$ for the mean of the underlying distribution is the *weighted average* defined as

$$\langle x \rangle = \frac{1}{w} \sum_{i=1}^{n} w_i x_i; \quad w = \sum_{i=1}^{n} w_i, \tag{5.18}$$

where the *weight factors* w_i are proportional to $1/\sigma_i^2$. Only proportionality is needed because the sum is divided by the total weight. Why this is the correct way of averaging is explained in Appendix A8 on page 158.

This type of averaging does not only apply to x but to any quantity that is to be averaged, e.g.

$$\langle x^2 \rangle = \frac{1}{w} \sum_{i=1}^{n} w_i x_i^2; \quad w = \sum_{i=1}^{n} w_i, \tag{5.19}$$

or, in general,

$$\langle f(x) \rangle = \frac{1}{w} \sum_{i=1}^{n} w_i f(x_i); \quad w = \sum_{i=1}^{n} w_i. \tag{5.20}$$

Accuracy of the estimated mean

When the mean of a data series has been estimated by weighted averaging of $x_i \pm \sigma_i$, then the estimate for the standard inaccuracy of the estimated mean is given by

$$\hat{\sigma}_{\langle x \rangle} = \left(\sum_{i=1}^{n} \frac{1}{\sigma_i^2} \right)^{-1/2}. \tag{5.21}$$

Why this is so is also explained in Appendix A8. Using this formula we assume that the values of σ_i^2 are reliable; we have not used the value of $\langle (\Delta x)^2 \rangle$ for the estimation of $\hat{\sigma}_{\langle x \rangle}$. Whether the observed spread in the measured values will be statistically acceptable (i.e., compatible with the known σ_i^2), can be tested with a *chi-squared test*. The chi-squared test will be fully treated in Section 7.4 on page 95 (see also the data sheet chi-squared distribution on page 199), but here we already make superficial use of it. For this case the number of degrees of freedom equals $n - 1$ and χ^2 is defined as

$$\chi^2 = \sum_{i=1}^{n} \frac{(x_i - \langle x \rangle)^2}{\sigma_i^2} = \frac{\langle (\Delta x)^2 \rangle}{\hat{\sigma}_{\langle x \rangle}^2}. \tag{5.22}$$

Note that $\langle (\Delta x)^2 \rangle$ must have been determined by weighted averaging according to (5.20). In the last term we have used (5.21). The value of χ^2 should be in the neighborhood of the number of degrees of freedom $n - 1$. How much it can reasonably deviate from this value is given by the cumulative chi-squared distribution (see page 2 of the data sheet **chi-squared distribution** on page 199).

If the σ_i's are *not* accurately known *and* the number of observations is sufficiently large, it is possible to use $\langle (\Delta x)^2 \rangle$ for the determination of $\hat{\sigma}_{\langle x \rangle}$. In that case assume that $\chi^2 = n - 1$, so that

$$\hat{\sigma}_{\langle x \rangle}^2 = \frac{\langle (\Delta x)^2 \rangle}{n - 1}. \qquad (5.23)$$

This equation – as expected – also applies to the case of independent samples of equal weight, and is therefore equivalent to (5.13) on page 59.

Which choice of method you make is up to you. When your individual variance estimations are unreliable, choose the latter method. To be on the safe side, you may also choose the largest of the two uncertainties from the two methods.

5.7 Robust estimates

Estimates of parameters like standard deviation and standard error, as have been treated in the previous sections, are quite sensitive to outliers in the data. The reason for this is the use of squared deviations; an outlier contributes rather heavily to a sum of squares. When a deviation is so large that its occurrence in the data set is rather unlikely, one may eliminate such an observation (see below). Some of the methods treated in the previous sections are only valid for normally distributed data, such as the confidence intervals determined by Student's t method. In modern statistics *robust* methods have been developed to handle data series in such a way that outliers play a lesser role and the results depend less strongly on the type of distribution function of the data. These robust methods are based on the *ranking order* of the data ('rank-based methods'). In this book we give only a brief summary of these methods and refer for further details to the literature (Petruccelli *et al.*, 1999; Birkes and Dodge, 1993; Huber and Ronchetti, 2009).

Elimination of outliers

It may happen that a particular measured value falls outside the expected range of values. This may be due to a random fluctuation, but it can also be the result of an experimental error or mistake. It is warranted to eliminate such a data point from the data series before further processing. A reasonable, and often used, criterion is that the deviation exceeds 2.5 σ. Don't apply such elimination more than once in a data series. Prudence is required, because the choice whether or not to eliminate a data point may be influenced by subjective considerations if the particular measurement does or does not suit your purposes. Of course, rather than elimination, it is much better to repeat the measurement: a possible error or mistake can then be identified. If repeated measurements also show significant deviations from the expected value, you may be on the track of an interesting phenomenon worth further investigation.

The 2.5 σ criterion is rather arbitrary and many researchers prefer a limit of 3 σ. The criterion should be chosen such that the random occurrence of a value beyond the chosen limit is an unlikely event, e.g. with a probability of less than 5%. But with such a criterion the limiting value depends on the *number* of data points in the data series: the probability – given a normal distribution – that a *single* point deviates more than 2.5 σ is a bit over 1%; the probability that *at least one* point in a series of 20 data points deviates more than 2.5 σ exceeds 20%. The first case is unlikely, but the second case may easily occur by random sampling. In the table on page 2 of the data sheet NORMAL DISTRIBUTION on page 205 the probability is tabulated that *at least one* data point out of n points falls outside the range $(\mu - d, \mu + d)$ (a *two-sided* criterion), for various values of d/σ; on page 4 the probability is tabulated that at least one data point exceeds the value $\mu + d$ (a *one-sided* criterion). If you choose a 5% limit for this one-sided probability, you see that for less than 10 data points 2.5 σ is a good choice; for 10 to 50 points 3 σ is better and for 50 to several hundred points 3.5 σ is the best choice.

Rank-based estimates

The estimated mean of a distribution is usually taken as equal to the average of the measured values. When you have a good reason to assume a symmetric underlying probability distribution, but no good reason to assume a normal distribution, you can also take the *median* of the measured values. For a large number of points the result is the same, but for a small number of points the median is less sensitive to outliers than the average. The median has the property that the number of positive and negative deviations are equal; in order to obtain the median only the *sign* of the deviations is used.

Sign-based confidence intervals

A sign-based estimate of a confidence interval is obtained from the binomial distribution of the number of positive signs of the possible deviations. Suppose you have five measurements, sorted in ascending order: x_1, x_2, x_3, x_4, x_5. As estimate for the mean $\hat{\mu}$ you take the median x_3. Now consider the probability that the $\mu < x_1$. In that case the deviations would have the signs $+++++$ and the binomial probability of obtaining five pluses when each sign has a 50% chance of being plus, is

$$p(\mu < x_1) = 2^{-5} \binom{5}{5} = 1/32 \qquad (5.24)$$

(see Section 4.3). The same probability is obtained when $\mu > x_5$, so the interval (x_1, x_5) has a confidence level of $30/32 = 94\%$. When μ lies between x_2 and x_4, the deviations have the sign $--+++$ or $----++$; the binomial probability that this happens is:

$$p(x_2 < \mu < x_4) = 2^{-5} \binom{5}{3} + 2^{-5} \binom{5}{2} = 20/32 = 62\%. \qquad (5.25)$$

Because only a small number of discrete values are available, the interval for a preset confidence level of, say, 90%, cannot be given. The method is robust, but also quite inaccurate. If there is an indication that the data points are samples from a normal distribution, the "classical" parameter estimates are much better. In order to keep in line with the classical report of the standard deviation, an alternative robust estimate of the "standard deviation" can be obtained by abstracting the 68% confidence interval from an analysis of the cumulative distribution function (see Section 5.1 on page 54).[6]

The bootstrap method

Finally a few words about another *distribution-free* method, the *bootstrap*, designed to obtain an approximate probability distribution (a "sampling distribution") of an estimated mean on the basis of a data series, without any assumption about the probability distribution from which the data are sampled. The method originates from 1979, see Efron and Tibshirani (1993). The method is simple but can only be realized by using a computer.

Assume you have a data series of n independent samples of equal weight from an unknown distribution. The average of this series is a good estimate for the mean of the distribution. You wish to generate a number of such averages to produce a sampling distribution of the mean; this gives you details on the accuracy of the estimated mean. Unfortunately, this is only possible if you could produce many new sets of measurements, which would each freshly sample the data distribution. But you don't have more new data, so you must rely on the n samples you already have. Now generate a large number (say, 3000) of series, each consisting of n "measurements," each drawn randomly from the n original data, but with *replacement*, i.e., without changing the probability of drawing a particular value. From each series determine the average. The collection of all 3000 averages so obtained approximates

---

[6] The deviation d for which the interval $(\mu - d, \mu + d)$ equals the 68 percent confidence interval is strictly not a standard deviation and it does not imply the validity of other confidence intervals derived from normal distributions. It should be checked if this value equals the best estimate of the standard deviation within its error limits.

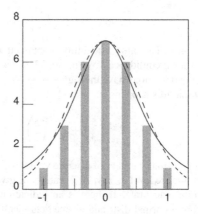

Figure 5.4 Histogram of the bootstrap distribution of the mean of three data values −1, 0 and 1 (plotted ordinates ×1/27). Drawn line: Student's t-distribution for two degrees of freedom; dashed line: normal distribution with "classical" standard deviation. All distributions have been scaled to yield the same maximum.

the true sampling distribution you would have obtained from 3000 sets of fresh measurements.

For a small number of measurements it is possible to generate *all* possible series (there are n^n of those), but for more than five data points this runs out of hand. For illustrative purposes Fig. 5.4 gives the bootstrap distribution for three data points with values −1, 0 and 1: there are seven possible averages. In the same figure the Student's t-distribution is given for the same three measured values (i.e., for two degrees of freedom), for which $\hat{\mu} = 0$ and $\hat{\sigma} = 1$. The "classical" standard uncertainty of the mean equals $\hat{\sigma}/\sqrt{3} = 0.577$; the s.d. of the bootstrap distribution is $\sqrt{2}/3 = 0.471$. The latter is also the standard uncertainty of the mean using the biased estimator: $\sqrt{\langle (\Delta x)^2 \rangle}/\sqrt{3}$. Also the normal distribution with $\sigma = 0.577$ is given. We see that for this symmetric case the normal distribution and the bootstrap distribution agree well; the t-distribution has broader flanks. If you only have three values and no good reason to assume normality of the underlying distribution, there is no good reason to apply the Student's t-distribution.

The term "bootstrap" now becomes meaningful: A bootstrap method is a method to obtain something new from nothing, which in principle is impossible, such as lifting yourself off the ground by pulling your bootstraps. Have we gained anything new by applying the bootstrap method? No! The bootstrap produces an array of averages of n samples taken from a given distribution: a sum of n δ-functions at the original data points. The distribution function of these averages can be computed by methods treated in Appendix A5; its mean and standard deviation are completely determined by the original data. In fact, the mean of the bootstrap distribution equals the mean of the original data and the s.d. of the bootstrap distribution equals the rmsd $\sqrt{\langle(\Delta x)^2\rangle}$ of the original data divided by $\sqrt{n}$. This equals the *biased* estimate of the standard uncertainty in the mean; we know that the *unbiased* estimate rmsd divided by $\sqrt{n-1}$ is better. A bootstrap distribution with unbiased s.d. can be obtained by adding $n-1$ rather than n samples from the original data.

So it seems that the bootstrap method is rather meaningless. It is meaningless, indeed, for obtaining a best estimate for the mean and its standard uncertainty. It is not meaningless, however, for obtaining confidence intervals for a given confidence level. But you should always be aware of the fact that the bootstrap distribution does not extend beyond the minimum and maximum data value, while the underlying probability distribution may well have tails extending (far) beyond those values. Confidence limits derived from the tails of the bootstrap distribution may well be unrealistically narrow and could lead to erroneous conclusions. See Exercise 5.6 for a comparison of various estimates.

A program that will generate an array with averages from random samples taken from a given dataset is **Python code** 5.1 on page 175.

A program `report` to analyze a data set is **Python code** 5.2 on page 176. Given a set of independent data, it produces a graph of the cumulative distribution (on a probability scale) and a graph with data points and standard deviations (if given); it prints properties of the data (including skewness and excess) and identifies outliers. In addition it performs functions explained in Chapter 7: drift analysis with significance tests and a chi-squared analysis if standard deviations are given. Look for updates on www.hjcb.nl.

Summary *You are now able to make a clear distinction between the distribution of measured data x_i and the (unknown) underlying distribution from which the data are supposed to be random samples. Properties of your measured data are number n, average $\langle x\rangle$, mean squared deviation (msd) $\langle(\Delta x)^2\rangle$, and root-mean-squared deviation (rmsd) $\sqrt{\langle(\Delta x)^2\rangle}$, but also rank-based properties such as range, median and various percentiles. From these properties you can derive best estimates $\hat{\mu}$, $\hat{\sigma}$ for the parameters of the underlying distribution: mean and standard deviation. An important quantity is the inaccuracy of the estimated mean $\sigma_{\langle x\rangle}$ (the s.d. of the sampling mean), which equals*

$\hat{\sigma}/\sqrt{n}$. *You are aware of the fact that all these formulas are valid for a set of n independent samples; if samples are correlated, the estimated variance becomes somewhat $(((n-1)/(n-n_c))\times)$ larger and the standard inaccuracy in the sample mean becomes considerably $(n_c \times)$ larger, where n_c is a correlation length. You know how to handle your data if the data points have unequal statistical weights: in all kinds of averaging, the values to be averaged are multiplied by their weight w_i/w, where w is the total weight.*

You can express results in terms of confidence intervals. These can be one-sided or double-sided. For example, a 90 percent double-sided confidence interval gives the estimated range from the 5th to the 95th percentile of the underlying distribution. For normally distributed variables, the confidence intervals follow from the normal distribution if you know σ beforehand, or from Student's t-distribution if you don't. An alternative determination of confidence intervals for the sampling mean is to construct a bootstrap distribution based on your data itself. You are aware of the pitfalls of this "distribution-free" method.

Finally, if you have a set of data and a good prior estimate of the inaccuracy of each data point, you can use the chi-squared distribution to assess whether the spread in the measured data is compatible with the a-priori inaccuracies. If the spread is improbably large, there is probably an error source that you overlooked.

Exercises

5.1 Could the data given in Table 2.1 on page 6 be sampled from a normal distribution? If so, estimate $\hat{\mu}$ and $\hat{\sigma}$ by drawing a straight line through the cumulative distribution function of Fig. 2.1.

5.2 Prove (5.6).

5.3 If you subtract a constant from all values of x and then compute the msd using (5.6), is a further correction still required?

5.4 Generate 1000 normally distributed variables with mean c and s.d. 1. Compare the rmsd computed by both (5.4) and (5.6). Vary the constant c (e.g. 1.e6, 1.e7, 1.e8, 1.e9).

5.5 (refer to Table 5.1 on page 61)
A series of n independent measurements of a physical quantity yields an average of 75.325 78 and a mean squared deviation of 25.643 06. Report,

with the correct number of digits, your best estimates of the mean and standard deviation of the underlying probability distribution, for two cases: (a) $n = 15$, (b) $n = 200$.

5.6 You live in Germany and want to calibrate the speedometer of your car. On a quiet, mostly straight and level Autobahn section you keep your speed as accurate as possible at 130 km/hr on your speedometer. Your companion measures with a stopwatch the time between passing two kilometer marks that are exactly 1 km apart. She finds the following nine intervals (in s):[7]
29.04, 29.02, 29.24, 28.89, 29.33, 29.35, 29.00, 29.25, 29.43

1. Compute the following properties of the measured set of time intervals:
    (a) the average,
    (b) the average squared deviation from the average,
    (c) the root-mean-squared average deviation from the average,
    (d) the range, median and the first and third quartiles.

2. Compute the best estimates for the following properties of the underlying distribution function:
    (a) the mean $\hat{\mu}$,
    (b) the variance $\hat{\sigma}^2$,
    (c) the standard deviation $\hat{\sigma}$,
    (d) the standard uncertainty of the estimated mean,
    (e) the uncertainty of the last three values.

3. What is (the best estimate for) your car's real velocity? What is the standard uncertainty of this value? How large is the speedometer's deviation and what is the relative accuracy of that deviation? Give all values with the correct number of significant digits.

4. If you as driver assert that you have kept the speed within a deviation of ±0.5 km/hr, does this knowledge influence your conclusions in any way?

5. Assuming that the (biased) bootstrap yields a reliable sampling distribution of the mean, generate a bootstrap distribution of 2000 samples and compute the 80%, 90% and 95% confidence limits for the time interval.

6. Using this bootstrap distribution, compute the 80%, 90% and 95% confidence limits for the velocity.

7. Assuming the underlying distribution to be normal $N(\hat{\mu}, \hat{\sigma})$, compute the 80%, 90% and 95% confidence limits for the velocity.

---

[7] These numbers are from a real experiment.

8. Assuming the underlying distribution to be normal with unknown s.d., compute the 80%, 90% and 95% confidence limits for the velocity according to Student's t-distribution.

5.7 You are a member of a CODATA committee with the task to update Avogadro's number. The following reliable data are at your disposal:

- the already known number (see data sheet PHYSICAL CONSTANTS on page 209)
- a series of measurements by scientist A with result:
  $6.022\,141\,48(75) \times 10^{23}$
- a series of measurements by scientist B with result:
  $6.022\,142\,05(30) \times 10^{23}$
- a series of measurements by scientist C with result:
  $6.022\,1420(12) \times 10^{23}$

Give the weighted mean and its standard uncertainty.

5.8 Plot the bootstrap distribution, the histogram of which is given in Fig. 5.4, on a probability scale. Is this distribution compatible with a normal distribution? Estimate graphically the mean and s.d. and compare to the values given in the text.

5.9 (*This advanced exercise requires reading of Appendix A3 and Appendix A5.*)
Determine – using the characteristic function – the distribution function of the sum of three samples, each randomly chosen with equal probability from the three values -1, 0 and 1. Note that the distribution function for the sum of three values equals the convolution of the distribution functions of each value. Determine its variance. Compare your result with Fig. 5.4.

6 Graphical handling of data with errors

Often you perform a series of experiments in which you vary an independent variable, such as temperature. What you are really interested in is the relation between the measured values and the independent variables, but the trouble is that your experimental values contain statistical deviations. You may already have a theory about the form of this relation and use the experiment to derive the still unknown parameters. It can also happen that the experiment is used to validate the theory or to decide on a modification. In this chapter a global view is taken and functional relations are qualitatively evaluated using simple graphical presentations of the experimental data. The trick of transforming functional relations to a linear form allows quick graphical interpretations. Even the inaccuracies of the parameters can be graphically estimated. If you want accurate results, then skip to the next chapter.

6.1 Introduction

In the previous chapter you have learned how to handle a series of equivalent measurements that should have produced equal results if there had been no random deviations in the measured data. Very commonly, however, a quantity y_i is measured as a function $f(x_i)$ of an independent variable x_i such as time, temperature, distance, concentration or bin number. The measured quantity may also be a function of several such variables. Usually the independent variables – which are under the control of the experimenter – are known with high accuracy and the dependent variables – the measured values – are subject to random errors. In that case

$$y_i = f(x_i) + \varepsilon_i, \tag{6.1}$$

where x_i is the independent variable (or the set of independent variables) and ε_i is a random sample from a probability distribution.

Generally, you already have a theory about the function f, although that theory may contain unknown parameters $\theta_k (k = 1, \ldots, m)$:

$$y = f(x, \theta_1, \ldots, \theta_m). \tag{6.2}$$

An example is the linear relation

$$y = ax + b, \tag{6.3}$$

but the relation can be more complex like

$$y = c \exp(-kx). \tag{6.4}$$

It is often possible to *linearize* the relation by a simple transformation. For the latter case:

$$\ln y = \ln c - kx \tag{6.5}$$

yields a linear relation between $\ln y$ and x. It is usually recommended to make such a linearization, as a simple graphic plot will show a straight line, permitting a quick judgment of the suitability of your presumed functional relation. In Section 6.2 a few examples will be worked out.

Let us return to the linear relation $y = ax + b$. Suppose you measured n data points (x_i, y_i), $i = 1, \ldots, n$, and expect the measured values y_i to satisfy *as closely as possible* the relation

$$y_i \approx f(x_i), \tag{6.6}$$

where $f(x) = ax + b$ is the expected relation. Your task is to determine the parameters a and b such that the measured values y_i deviate as little as possible from the function values. But what does that mean? The deviations ε_i of the measured values with respect to the function:

$$\varepsilon_i = y_i - f(x_i) \tag{6.7}$$

should be the sole consequence of random errors and we expect in general that the deviations ε_i are random samples from a probability distribution with zero mean. In practice this distribution is often normal. The correct method for this kind of parameter estimation is the *least-squares fit*, which is treated in Chapter 7. A computer program is needed to perform a least-squares fit.

It is not always necessary to perform a precise least-squares fit. It is always meaningful to plot the data in such a way that you expect a linear relation. A straight line can be adequately judged by visual inspection. A straight line drawn "by eye" to fit the points often gives sufficiently accurate results and even the inaccuracies in the parameters a and b can be estimated by varying the line within the cloud of measured data points. There is nothing wrong with making a quick sketch on old-fashioned graph paper! Computer programs are useful when there are many data points, when different points have different weights or when high accuracy is required, but they are never a substitute for bad measurements and almost never give you more insight into the functional relations. Be careful with computer programs that are not well-documented or do something you don't quite understand!

This chapter is devoted to simple graphical processing of experimental data with a simple discussion of the inaccuracies in the results. Always ask yourself if such a simple graphic analysis can be useful for your problem: often you get a better insight into the relation between model and data. After having done a simple analysis, a more accurate and elaborate computer analysis can (and should) be made.

6.2 Linearization of functions

In this section a few examples of the linearization of functions are given.

(i) $y = ae^{-kx}$: $\ln y = \ln a - kx$ (examples: concentration as function of time for a first-order reaction, number of counts per minute for a radioactive decay process). Plot $\ln y$ on a linear scale versus x, or plot y on a logarithmic scale versus x. If you do this by hand, use semi-log paper (one coordinate linear, the other logarithmic with e.g. two decades). Or use a simple Python plot. Figure 2.7 on page 16 is an example. The slope ($-k$ in this example) is read from the graph by selecting a segment (take a large segment for better accuracy) and read the coordinates of the end points (x_1, y_1) and (x_2, y_2); the slope equals $\ln(y_2/y_1)/(x_2 - x_1)$. If you take a full decade for the end points (e.g. passing through $y = 1$ and $y = 10$), then the slope is simply $\ln 10/(x_2 - x_1)$.

(ii) $y = a + be^{-kx}$: $\ln(y - a) = \ln b - kx$. First estimate a from the values of y for large x and then plot $y - a$ versus x on a logarithmic scale. If the plot doesn't yield a linear relation, adjust a somewhat (within reasonable bounds).

(iii) $y = a_1 e^{-k_1 x} + a_2 e^{-k_2 x}$. This is difficult to handle graphically, unless k_1 and k_2 are very different. A computer program also has difficulties with this kind of analysis! First estimate the "slow" component (with smallest k), subtract that component from y and plot the difference on a logarithmic scale. Figure 6.1 gives the result for the data given in Table 6.1; the standard error in each y is ± 1 unit.

The column z in Table 6.1 gives the differences between y and the values given by the line in the left panel of Fig. 6.1. This line has been drawn "by eye" and goes through the points $(0, 25)$ and $(100, 2.5)$, yielding $k_2 = [\ln(25/2.5)]/100 = 0.023$. Hence the equation for this line is

Table 6.1 *Measured values y that result from a sum of*
two exponentials. The column z results from subtraction
of the "slowest" exponential. The standard uncertainty
in y equals one unit.

x	y	z	x	y	z
0	90.2	65.2	40	11.7	1.7
5	62.2	39.9	50	8.8	0.9
10	42.7	22.9	60	6.9	0.6
15	30.1	12.4	70	4.6	−0.4
20	23.6	7.8	80	5.0	1.1
25	17.9	3.8	90	2.9	−0.3
30	14.0	1.5			

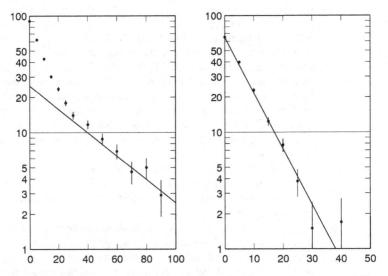

Figure 6.1 Graphical analysis of data which represent the sum of two exponentially
decaying quantities. In the left panel the data points y have been plotted on a logarithmic scale versus the independent variable x and the "slowest" component is
approximated by a straight line. In the right panel the differences z between the data
y and the "slow" component are plotted. Note the different scales for x.

25 exp($-0.023x$). In the right panel of this figure z has been plotted: the points approximately follow a linear relation. The drawn line goes through the points $(0,65)$ and $(38, 1)$, yielding $k_1 = (\ln 65)/38 = 0.11$. Therefore, the function that approximates the behavior of all data points is given by

$$f(x) = 65\,e^{-0.11\,x} + 25\,e^{-0.023\,x}. \tag{6.8}$$

This simple graphical approach does not provide a solid basis to make a reliable guess of the uncertainties in the parameters of this equation. But it provides an excellent basis for the *initial guess* of the parameters in a *nonlinear least squares fit*. The latter is the subject of Chapter 7. Such a fit must be carried out by computer; a suitable program not only provides the best fit, but also gives an estimate of the inaccuracies and correlations of the parameters.

(iv) $y = (x - a)^p$ (example: the isothermal compressibility χ of a fluid in the neighborhood of the critical temperature behaves as a function of temperature according to $\chi = C(T - T_c)^{-\gamma}$, where γ is the *critical exponent*). Plot $\log y$ versus $\log(x - a)$ (or y versus $(x - a)$ on a double-logarithmic scale); if a is not known beforehand, then vary a somewhat until the relationship becomes a straight line. The slope of the line yields p.

(v) $y = ax/(b + x)$ (examples: adsorbed quantity n_{ads} of a solute versus concentration c in solution or versus pressure p in the gas phase in the case of Langmuir-type adsorption: $n_{ads} = n_{max}c/(K + c)$; reaction rate v as function of substrate concentration $[S]$ in the case of Michaelis–Menten kinetics[1] $v = v_{max}[S]/(K_m + [S])$). By taking the reciprocal of both sides, this equation becomes a linear relation between $1/y$ and $1/x$:

$$\frac{1}{y} = \frac{1}{a} + \frac{b}{a}\frac{1}{x}. \tag{6.9}$$

In enzyme kinetics a graph of $1/v$ versus $1/[S]$ is called a *Lineweaver–Burk plot*.[2] There are two other ways to produce a linear relation: plot x/y versus x (the *Hanes method*):

$$\frac{x}{y} = \frac{b}{a} + \frac{x}{a}, \tag{6.10}$$

or plot y/x versus y (the *Eadie–Hofstee method*):

$$\frac{y}{x} = \frac{a}{b} - \frac{y}{b} \tag{6.11}$$

---

[1] This will be familiar to you if you are a biochemist, but sound as abacadabra if you are a physicist or mechanical engineer. You may consult any textbook on biochemistry for details. Or think of an application in your own field that leads to this kind of equation.

[2] See e.g. Price and Dwek (1979).

Table 6.2 *Conversion rate v of urea by the enzyme urease as function of the urea concentration [S]. The reciprocal values are given to produce a Lineweaver–Burk plot. The standard uncertainty in $1/v$ equals σ_v/v^2.*

[S] mM	$1/[S]$ mM^{-1}	v $mmol\,min^{-1}$	σ_v mg^{-1}	$1/v$ $mmol^{-1}\,min$	$\sigma_{1/v}$ mg
30	0.03333	3.09	0.2	0.3236	0.0209
60	0.01667	5.52	0.2	0.1812	0.0066
100	0.01000	7.59	0.2	0.1318	0.0035
150	0.00667	8.72	0.2	0.1147	0.0026
250	0.00400	10.69	0.2	0.09355	0.0018
400	0.00250	12.34	0.2	0.08104	0.0013

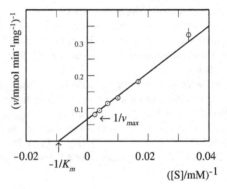

Figure 6.2 Lineweaver–Burk plot of the tabulated data.

Which method to choose depends on the inaccuracies of the data points: whenever a reciprocal of x or y is used, small values get relatively more importance in the plot.

Example: urease kinetics

With the experimental values for the rate of conversion $v = y$ of urea by the enzyme urease[3] as a function of the urea concentration [S]$= x$ as given in Table 6.2, the plots of Fig. 6.2 and Fig. 6.3 are obtained. In a Lineweaver–Burk plot the value of $K_m = b$ can be obtained from the intersection with

---

[3] Example taken from Price and Dwek (1979), with additional noise.

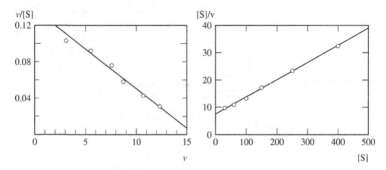

Figure 6.3 Eadie–Hofstee (left) and Hanes (right) plot of the tabulated data.

the horizontal (x) axis and the value of $v_{max} = a$ can be obtained from the intersection with the vertical (y) axis. The estimation of inaccuracies of the parameters from these graphs is not reliable; also in this case it is better to perform a nonlinear least-squares analysis using the graphical estimates for the initial guess of the parameters.

6.3 Graphical estimates of the accuracy of parameters

In the previous section you have seen how you can plot your data in such a way that a linear relationship is obtained and how you can estimate the two parameters of a linear function by drawing the "best" line through the data points. In this section you will see how you can make a simple estimate of the uncertainties in those parameters. Sometimes such estimates are sufficient. If they are not, a more accurate *least-squares fit* is required.

In order to be able to estimate the uncertainties, you need to include error bars in the graphs. When the uncertainty in the independent variable x, plotted on the horizontal scale, is negligible, it suffices to use vertical error bars from $y - \sigma_y$ to $y + \sigma_y$. When there is a sizeable uncertainty in x, a horizontal error bar from $x - \sigma_x$ to $x + \sigma_x$ must be included as well. A clear presentation is an *ellipse* with principal axes of length $2\sigma_x$ and $2\sigma_y$.

The best straight line through the data points fits as closely as possible to all (x_i, y_i). The first requirement is that the line be drawn such that the sum of the deviations (sign included) is (close to) zero. But that does not determine the line! Any line through the "center of mass"[4] of the points ($\langle x \rangle, \langle y \rangle$) fulfills this criterion. We need this criterium to be fulfilled not only globally, but also locally. A good guess is the line constructed through *two* centers of mass, each of a group of data points, see Fig. 6.4.

---

[4] "Mass" is to be interpreted as "statistical weight."

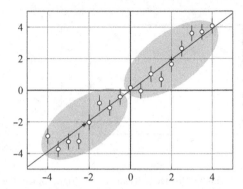

Figure 6.4 A line drawn through two "centers of mass" of two clouds of points approximates a linear fit to all points.

After drawing a straight line $f(x) = ax + b$, the parameters a and b can be determined from the slope and the value at which the line intersects the y-axis. The latter may be difficult to determine when the value $x = 0$ is outside the range of x-values of the data points. A much better method is to determine the "center of mass" $(\langle x \rangle, \langle y \rangle)$ of the points. The best fit should go through this point, as you shall see in Chapter 7. Only the slope a needs to be estimated:

$$f(x) = a(x - \langle x \rangle) + b, \qquad\qquad (6.12)$$

$$b = \langle y \rangle. \qquad\qquad (6.13)$$

The use of this relation has the advantage that uncertainties in the slope and the additive constant are uncorrelated with each other (see page 90). It is now much easier to estimate the uncertainties in a and b.

In order to estimate the uncertainties in the parameters, the line can be varied in slope a (Fig. 6.5) or in additive constant b (Fig. 6.6). As we know from the properties of a normal distribution, about 2/3 of the points should remain within the lines if a parameter is varied by $\pm\sigma$. So, as a rule of thumb, vary the parameters (one at a time) symmetrically such that about 15 percent of the points fall outside the lines on each side. Be aware of possible outliers that deviate conspicuously from the line. How to handle outliers has been treated in Section 5.7 on page 63: either eliminate or measure again!

6.4 Using calibration

Suppose you work with an instrument or method that produces a *reading y* (e.g. a digital number, a needle deflection, a meniscus height) from which

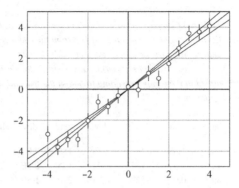

Figure 6.5 Linear fit through "center of mass" with slope varied by ±10% ($a = 1.0 \pm 0.1$).

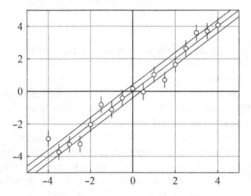

Figure 6.6 Linear fit through "center of mass" with additive constant varied by ±0.4 ($b = 0.0 \pm 0.4$).

a quantity x (e.g. a concentration, an electrical current, a pressure) must be deduced. When the instrument is not properly calibrated, i.e., when the reading does not correspond directly and reliably to the measured quantity, you should calibrate the instrument yourself. For this purpose you produce a *calibration table*, and preferably a *calibration curve*, by measuring the reading for a number of accurately known values of x. These data you either tabulate, or plot and interpolate in a curve, or express the relation between y and x in a mathematical function. Often you will tabulate a *correction table* or plot a *correction curve* that contains the differences between the readings and the correct values. Be sure to be explicit what the difference means: usually the correction is to be added to the reading to obtain the true value. In all cases you can deduce the value of x for any measured reading by inversion

Table 6.3 *Compass deviation chart of the U.S.S. Cleveland (1984)*
listing the compass deviations (dev) from the true magnetic
bearing for various ship headings (head). The deviation W (West)
means negative and E (East) means positive; the deviation is to be
added to the compass reading to obtain the true magnetic bearing
of the ship.

head	dev	head	dev	head	dev	head	dev
0	1.5W	90	1.0W	180	0.0	270	1.5E
15	0.5W	105	2.0W	195	0.5E	285	0.0
30	0.0	120	3.0W	210	1.5E	300	0.5W
45	0.0	135	2.5W	225	2.5E	315	2.0W
60	0.0	150	2.0W	240	2.0E	330	2.5W
75	0.5W	165	1.0W	255	2.5E	345	2.0W

of the calibration relation. How do you proceed and how do you determine
the uncertainty in x?
Be explicit!
Mariners and navigators have coped with magnetic compass corrections for
centuries, although modern electronic aids have diminished their problems.
The compass reading (C) must be corrected first by adding the *deviation*
due to the influence of magnetic and ferrous materials in the ship itself to
obtain the magnetic bearing (M); then the latter must be corrected by adding
the *variation* due to the position of the magnetic north pole – that does not
coincide with the true geometric north pole – to obtain the true bearing (T).
Traditionally deviation and variation are expressed as E (East) if positive,
or W (West) if negative. Since a sign error can have catastrophical consequences, sailors of all nations have invented mnemonics to remind them of
the proper sequence to add or subtract the corrections. A mannerly English
mnemonic is CADET: "**C**ompass **AD**d **E**ast (to get) **T**rue (bearing)", which
applies equally to deviation and variation. In the Dutch Navy Reserve (KMR)
the mnemonic "**K**ies **de M**eisjes **van R**otterdam" ("choose the girls of Rot-
terdam"): **K**ompas + **d**eviatie $\rightarrow$ **M**agnetisch + **v**ariatie $\rightarrow$ **R**echtwijzend
(True bearing) is more popular. But beware: American navigators reverse the
correction by the mnemonics "**T**rue **V**irgins **M**ake **D**ull **C**ompany" (True +
Variation $\rightarrow$ Magnetic + Deviation $\rightarrow$ Compass), which is wrong unless the
sign of the correction is also reversed. To remember this, they also memorize
"**A**dd **W**hiskey" to **A**dd **W**esterly corrections. So be careful and explicit in
all cases. See Table 6.3[5] and Fig. 6.7.

[5] Data from www.tpub.com/context/administration/14220/css/14220_64.htm.

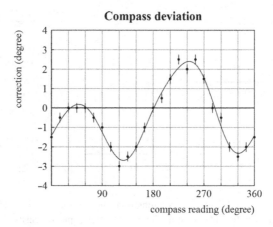

Figure 6.7 Graph of the compass deviations (Table 6.3). The error bars are ±0.25 degree, as the corrections are given with 0.5 degree precision. The drawn line is a least squares fit to a sum of Fourier components up to and including the fourth harmonic.

Python code 6.1 on page 182 shows how the least-squares Fourier components in Fig. 6.7 are computed. For general least square fits see Section 7.3.

Make sure in the calibration procedure that you cover the whole range of values for which the method will be used. Extrapolation is generally unreliable, but there is also no need to cover values that in practice will never occur. Draw the best line through the points; if the line is not straight, hopefully you can build it up from straight segments between calibrated points. If you want to be sophisticated, compute a *cubic spline* fitting function. Now, for any new measurement of x, given by a reading y, the quantity x can simply be read back from the calibration curve.

Now consider the inaccuracy of a measurement. There are two sources of error: one is the inaccuracy Δy in the reading y; the other is the inaccuracy in the calibration curve itself, due to inaccuracies of the calibration measurements. You should also be aware of additional errors that may occur, e.g. resulting from aging of the instrument after the last calibration. Both types of error lead to an uncertainty in x and both sources add up quadratically, because they are independent of each other. The two contributions are depicted in Fig. 6.8, which shows how a concentration in solution is deduced from a measurement of the optical density in a spectrometer. The

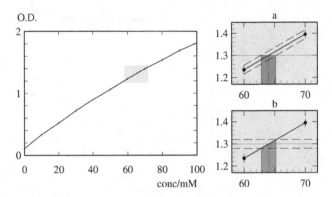

Figure 6.8 Example of a calibration line for spectrometric determination of the concentration of a chromophore in solution: optical density O.D. = log (*incident intensity/transmitted intensity*) as a function of the concentration of the solute. The gray area is magnified in the panels on the right: (a) the calibration error in the concentration, (b) the inaccuracy in the concentration resulting from the inaccuracy of the measured O.D.

calibration error is visualized by drawing two parallel sections of the calibration curve at distances representing the standard uncertainty in the calibration itself.

If the calibration has been very carefully performed, the calibration error is likely to be smaller than the direct error in the reading. In that case only the standard uncertainty σ_y of the reading counts. It leads to a standard uncertainty σ_x in the measured quantity by the relation

$$\sigma_x = \frac{\sigma_y}{\left|\left(\frac{dy}{dx}\right)_{cal}\right|}. \tag{6.14}$$

Summary *In this chapter you have learned how to plot your data in such a way that a functional relation becomes visible, preferably as a straight line. From simple plots you can roughly estimate the parameters of your function and – by varying the lines in position or slope – you can even get an idea of the inaccuracies of the parameters. You have also seen how calibrations are used to interpret instrument readings. You will not make errors in the sign when you apply calibrated corrections. The treatment in this chapter was rather sloppy, as its purpose was to provide a quick insight into your data. For more precision, proceed to the next chapter.*

Exercises

6.1 Draw a straight line "through" the points of Fig. 2.7 on page 16 and determine the parameters in $c(t) = c_0 \, e^{-kt}$.

6.2 From Figs. 6.2 and 6.3, determine the values of v_{max} and K_m. The straight lines drawn "by eye" go through the points $(-0.0094, 0)$ and $(0.04, 0.35)$ (Lineweaver–Burk), $(0.04, 0.35)$ and $(15, 0.007)$ (Eadie–Hofstee); $(0, 7.5)$ and $(500, 39)$ (Hanes).

6.3 Draw the best straight line through the data points of the logarithmic graph of k versus $1000/T$, made in Exercise 3.2 (page 25). Determine the constant E in the relation $k = A \exp(-E/RT)$ (which units?). Estimate the inaccuracy in E.

6.4 Using Fig. 6.8, determine the concentration (with s.d.) when the measured optical density equals 1.38 ± 0.01, assuming that the calibration error is negligible.

7 Fitting functions to data

If you want to fit parameters in a functional relation to experimental data, the best method is a least-squares analysis: Find the parameters that minimize the sum of squared deviations of the measured values from the values predicted by your function. In this chapter both linear and nonlinear least-squares fits are considered. It is explained how you can test the validity or effectiveness of the fit and how you can determine the expected inaccuracies in the optimal values of the parameters.

7.1 Introduction

Consider the following task: you wish to devise a function $y = f(x)$ such that this function fits as accurately as possible to a number of data points (x_i, y_i), $i = 1, \ldots, n$. Usually you have – based on theoretical considerations – a set of functions to choose from, and those functions may still contain one or more yet undetermined *parameters*. In order to select the "best" function and parameters you must use some kind of *measure* for the deviation of the data points from the function. If this deviation measure is a single value, you can then select the function that minimizes this deviation.

This task is not at all straightforward and you may be lured into pitfalls during the process. For example, your choice of functions and parameters may be so large and your set of data may be so small that you can choose a function that *exactly* fits your data. If you have n data points, you can fit an $(n - 1)$th degree polynomial exactly through all points. But you can also fit a smooth cubic spline through all points. In fact, there are an infinite number of functions that fit all points and by choosing one you have achieved nothing else than a fancy description of your data set (see Fig. 7.1). At best you have found a non-exclusive way to interpolate your data.

Two things are needed to improve the quality of your task. First, there must be a valid theory behind your choice of functions. The better your theory is, the more restricted is the range of functions and parameters from which you can choose. Second, your deviation measure must have statistical relevance

Fit nine points

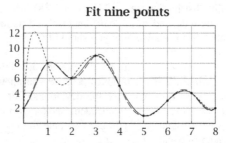

Figure 7.1 Several functions fitting exactly through nine equidistant points. The first and last point are taken equal to allow periodic solutions. Drawn curve: periodic cubic spline (piecewise third-degree polynomials with continuous first and second derivatives). Broken curve: each point is expanded by the function $\sin \pi x/(\pi x)$ (Nyquist–Shannon formula; the Fourier transform of the resulting function has no components with wavelength smaller than two units). Dotted curve: fit to a (non-periodic) eighth-degree polynomial using Lagrange's formula (Press *et al.*, 1992). Global polynomial fits are almost never satisfactory.

in the sense that it must have an associated *probability* that a given deviation occurs. For example, you have n independent data points ($n \gg 3$) and a theory that allows either a linear relation (two parameters) or a quadratic relation (three parameters). It is clear that a quadratic relation (which includes linear relations as a subset) will always fit better than a linear relation and a deviation measure will then be lower for the quadratic equation. Without a probability attached to your deviation measure you may always decide in favor of the quadratic relation, but with a proper probability measure you may decide that the quadratic relation fits too well and the linear relation is more probable. As you will see, with certain assumptions deviation measures can be devised to which meaningful probabilities can be attached.

Assuming that the independent variables x_i are accurate, and the function $f(x)$ is the correct relation, you expect that the deviations of the dependent variables y_i from the function values $f_i = f(x_i)$ behave as independent random samples from a probability distribution with zero mean and finite variance:

$$y_i = f(x_i) + \varepsilon_i; \tag{7.1}$$

$$E[\varepsilon_i] = 0, \tag{7.2}$$

$$E[\varepsilon_i \varepsilon_j] = \sigma_i^2 \delta_{ij}. \tag{7.3}$$

The ε_i's are called the *residuals* of the fitting procedure. The assumption that the x_i's are accurate is for convenience only; in Section 7.2 it is shown how you should treat your data when the x_i's are themselves samples of a probability distribution (see (7.11) on page 88). Also the assumption that the residues are independent (at least: uncorrelated) is for convenience only: in Appendix A9 it is shown how you should treat your data if the residues are correlated.

If you are lucky you may have some advance information on the probability distribution of the residues because you know something about the random process that generates the deviations. For example, if you know that the residue ε_i is an independent sample from a normal distribution with known variance σ_i^2, you can estimate the probability that the set of (independent) residues $\varepsilon_1, \ldots, \varepsilon_n$ occurs:

$$P(\varepsilon_1, \ldots, \varepsilon_n) = \Pi_{i=1}^n p(\varepsilon_i) \propto \exp\left[-\sum_{i=1}^n \frac{\varepsilon_i^2}{2\sigma_i^2}\right] = \exp\left[-\frac{1}{2}\chi^2\right], \quad (7.4)$$

where χ^2 is defined as the weighted sum of square deviations:

$$\chi^2 = \sum_{i=1}^n \frac{(y_i - f(x_i))^2}{\sigma_i^2}. \quad (7.5)$$

This probability product can be considered as the *likelihood* of the fit: a function with a higher likelihood is more likely to occur and the best fit can be considered to be the one that *minimizes* χ^2. Having found the best fit, you can use the minimum value of χ^2 to assess the quality of your fit by performing a *chi-squared analysis*. This will be considered in Section 7.4. Also the inaccuracy of the parameters in the function (their variances and covariances) can be derived (Section 7.5). In Chapter 8 the principles behind choosing the "best" function are considered more carefully.

In practice you may not be so lucky as to know beforehand what random process generates the deviations. Often you don't know the individual variances, but you do know the *relative weights* w_i of the deviations. For example, if data point i was the average of 100 measurements and data point j was the average of 25 similar measurements, then point i should be given a relative weight 4 times larger than point j. Or, if you have a series of measurements t_i with similar uncertainty, but you use $y_i = \log t_i$ for your fitting procedure, you should give the value y_i a weight proportional to t_i^2. Refer to Exercise 7.6 on page 109 for an explanation. Instead of minimizing χ^2 you can now minimize the *weighted sum of square residues S*:

$$S = \sum_{i=1}^n w_i (y_i - f_i)^2, \quad (7.6)$$

but – of course – you can no longer use the minimal value of S to assess the quality of the fit. If you can trust the functional form, and you have reasons to *assume* that the residues are just random samples from a distribution with unknown variance, you can *derive* an estimate of the variance of the distribution. From that you can in turn derive the inaccuracy (variances and covariances) of the parameters in the function.

So, for the determination of accuracies in a fitting procedure there are two possibilities: either use the known uncertainties in the data (if available) or use the observed magnitude of the sum of square deviations. If both are compatible, use the more reliable of the two, or – in doubt – choose the higher uncertainty. If they are both reliable but not compatible (according to a chi-squared analysis), then lean back, check your data and your error estimates – possibly measure again – and revise your theory.

7.2 Linear regression

Linear regression is a least-squares fit of the parameters in a *linear* function to a data set:

$$f(x) = ax + b, \tag{7.7}$$

where a and b are the adjustable parameters of the function. Given a set of independent data (x_i, y_i), $i = 1, \ldots, n$, with optional individual weights w_i, it is now your task to minimize the sum of (weighted) square deviations:

$$S = \sum_{i=1}^{n} w_i (y_i - f_i)^2 \text{ minimal} \tag{7.8}$$

by adjusting the parameters a and b. Here

$$f_i = f(x_i) = ax_i + b. \tag{7.9}$$

Here we use x to denote the independent variable, which is also called the *explanatory* variable, as the values of y – but for a random deviation – follow from x. There may well be more than one explanatory variable so that $f_i = f(\mathbf{x})$, where $\mathbf{x}$ is a vector and also the parameter a becomes a vector. This complicates the least-squares solution a bit; the multidimensional linear regression can be found in Appendix A9.2 on page 161.

While (7.7) is linear in x, it is the *linearity in the parameters a* and b that allows an analytical solution of the minimization problem (7.8). Thus also for functions like $ax^2 + bx + c$ or $a + b \log x + c/x$ least-squares minimizations can be solved by linear regression. How this is done is explained in Appendix A9. Here we consider only linear functions of x.

The factors w_i are the statistical weights of the data points. It is quite common that all points have the same weight because they come from the same statistical distribution; in those cases w_i can all be taken $= 1$. If the weights are not equal because the data points have different standard deviations σ_i, then the weights must be taken equal to (or proportional to) $1/\sigma_i^2$ (note: *not* proportional to $1/\sigma$!).

Uncertainties in x

When the uncertainties in x are negligible (which is the common case), the standard deviation σ_i is simply the standard deviation of y_i. When the uncertainty in x is not negligible (but independent of the deviation of y_i), then σ_i^2 must be replaced by

$$\sigma_i^2 = \sigma_{yi}^2 + \left(\frac{\partial f}{\partial x}\right)_{x=x_i}^2 \sigma_{xi}^2, \tag{7.10}$$

because we deal with the uncertainty in $y_i - f(x_i)$. For the linear relation (7.7) this reduces to

$$\sigma_i^2 = \sigma_{yi}^2 + a^2 \sigma_{xi}^2. \tag{7.11}$$

To evaluate this, you need to know the value of a which is yet to be determined. However, a rough estimate (e.g. from a graphic sketch) suffices at this point.

The best parameter estimates

In general you will need a computer program to find the solution to the least-squares minimization problem of (7.8). However, for the linear relation (7.7) the solution can be expressed in simple terms. The solution follows from setting the two derivatives $\partial S/\partial a$ and $\partial S/\partial b$ to zero and is worked out in Appendix A9. Here the resulting equations are given.

The parameters a and b follow from a number of averages over the measured data points. The weights must be taken into account to determine the averages, just as was done in Section 5.6 (see, for example, (5.20) on page 62). For example:

$$\langle xy \rangle = \frac{1}{w} \sum_{i=1}^{n} w_i x_i y_i; \quad w = \sum_{i=1}^{n} w_i. \tag{7.12}$$

The parameters are:

$$a = \frac{\langle (\Delta x)(\Delta y) \rangle}{\langle (\Delta x)^2 \rangle}; \quad b = \langle y \rangle - a\langle x \rangle, \tag{7.13}$$

where

$$\Delta x = x - \langle x \rangle; \quad \Delta y = y - \langle y \rangle. \tag{7.14}$$

These averages can also be computed without first subtracting the averages of x and y because

$$\langle (\Delta x)(\Delta y) \rangle = \langle xy \rangle - \langle x \rangle \langle y \rangle; \tag{7.15}$$

$$\langle (\Delta x)^2 \rangle = \langle x^2 \rangle - \langle x \rangle^2. \tag{7.16}$$

Beware of numerical precision when you subtract two large numbers (see note on page 58).

From the equation for b (7.13) you see that the optimal function passes through the point $(\langle x \rangle, \langle y \rangle)$. This is the "center of mass" of the set of points. This fact was used in the discussion of graphical estimates in Section 6.3.

Uncertainties in the parameters

Estimates for the standard uncertainties σ_a and σ_b in a and b are the square root of the estimated variances $\hat{\sigma}_a^2$ and $\hat{\sigma}_b^2$. The latter follow from the behavior of the function $\chi^2(a, b)$ and are derived from the likelihood function (7.4):

$$p(a, b) \propto \exp\left(-\frac{1}{2}\chi^2(a, b) \right). \tag{7.17}$$

The function $\chi^2(a, b)$ is a quadratic function in a and b and hence the probability distribution $p(a, b)$ is a bivariate normal distribution in the deviations Δa and Δb from the parameter values at the minimum. The coefficients of the terms $(\Delta a)^2$, $(\Delta a)(\Delta b)$ and $(\Delta b)^2$ determine the variances and covariance of a and b as explained in Section 7.5 and in Appendix A9. When χ^2 is estimated from the data themselves, i.e. from the minimum value S_0 of S, the results for the estimated (co)variances are

$$\text{var}\,(a) = \hat{\sigma}_a^2 = \frac{S_0}{w(n-2)\langle (\Delta x)^2 \rangle}, \tag{7.18}$$

$$\text{var}\,(b) = \hat{\sigma}_b^2 = \hat{\sigma}_a^2 \langle x^2 \rangle, \tag{7.19}$$

$$\text{cov}\,(a, b) = -\hat{\sigma}_a^2 \langle x \rangle \tag{7.20}$$

where w is the total weight of all points together. In the common case that all weights have been taken $= 1$, w is simply equal to the number of observations n.

The $n - 2$ in (7.18) has the meaning of the number of *degrees of freedom*: the number of (independent) data points minus the number of parameters in the function. Appendix A9 explains the details, but one could loosely say that two points are needed to determine two parameters and only $n - 2$ points are left to determine deviations from the fit. This makes sense: you can always draw a straight line through two points; for $n = 2, S = 0$ and the inaccuracies remain undetermined.

Covariances between parameters

The covariance $\mathrm{cov}\,(a, b)$ indicates whether deviations in a and b are correlated with each other: to what extent can a deviation in a be compensated by a deviation in b? Covariances must be used to determine the uncertainty in quantities that are a function of the parameters, e.g. for interpolation and extrapolation of data. See the remarks on page 22 and in Appendix A1.

The covariance is often expressed relative to the product $\hat{\sigma}_a \hat{\sigma}_b$ and is then called the *correlation coefficient* between a and b (a dimensionless number between -1 and $+1$):

$$\rho_{ab} = \frac{\mathrm{cov}\,(a, b)}{\hat{\sigma}_a \hat{\sigma}_b} = -\frac{\langle x \rangle}{\sqrt{\langle x^2 \rangle}}. \tag{7.21}$$

Note that a and b are uncorrelated ($\rho_{ab} = 0$) when $\langle x \rangle = 0$, i.e., when the zero of x is chosen in the "center of mass." So when you choose as the linear function

$$f(x) = a(x - \langle x \rangle) + b, \tag{7.22}$$

you can be sure that a and b are uncorrelated. In addition, you know right away that

$$b = \langle y \rangle. \tag{7.23}$$

Extrapolations are now much simplified: if you wish to determine the inaccuracy in $f(x)$ at an arbitrary x you can simply add the contributions quadratically:

$$\sigma_f^2 = \sigma_a^2 (x - \langle x \rangle)^2 + \sigma_b^2, \tag{7.24}$$

while using the formula $f(x) = ax + b$ a correction is needed: (see Appendix A1):

$$\sigma_f^2 = \sigma_a^2 x^2 + \sigma_b^2 + 2\rho_{ab}\sigma_a\sigma_b x. \tag{7.25}$$

Should you use S_0 or χ_0^2?

As you have noticed, the (co)variances are proportional to the minimal sum of squares: we have used the measured deviations to determine the uncertainties in the parameters. This is the only choice we have if the individual standard deviations σ_i are not known beforehand. If they *are* known and reliable, they could also be used to determine the uncertainties in the parameters. In that case the term $S_0/[w(n - 2)]$ must be replaced by the known

$1/\sum \sigma_i^{-2}$. Before deciding which choice to make you should always perform a chi-squared analysis (Section 7.4). These matters are discussed more fully in Sections 7.4 and 7.5.

Correlation coefficient between *x* and *y* values of a data series

There is a quantity that indicates how well a series of points lie on a straight line. This is the *correlation coefficient r* between the *x* and *y* values of the data series. Points approach a straight line only when there is a strong correlation between their *x* and *y* values. Don't confuse this correlation coefficient with the correlation coefficient ρ_{ab} (7.21) between *a* and *b* as discussed above. While the latter is derived from expectations over an estimated probability distribution, *r* is a property of the data set itself:

$$r = \frac{\langle (\Delta x)(\Delta y) \rangle}{\sqrt{\langle (\Delta x)^2 \rangle \langle (\Delta y)^2 \rangle}} \tag{7.26}$$

$$= a\sqrt{\frac{\langle (\Delta x)^2 \rangle}{\langle (\Delta y)^2 \rangle}}. \tag{7.27}$$

For $r = \pm 1$ there is complete correlation: the points lie exactly on a straight line; for $r = 0$ there is no correlation and it makes no sense to fit a linear function to the data. For a reasonable correlation *r* should be above 0.9.

A correlation coefficient less than 1.0 indicates *that*, but not *how* the data deviate from a linear relation. The two sets of data plotted in Figs 7.2a and 7.2b have the same correlation coefficient of 0.900, but both deviate in very

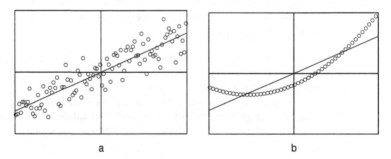

a   b

Figure 7.2 Two data sets x_i, y_i with the same correlation coefficient *r* between *x* and *y* of 0.900. There are no numbers along the axes: correlation coefficients are independent of linear scaling or translation of the axes. The drawn lines are the best linear fits through the points.

different ways from the best-fitted straight line. You also see in this figure that $r = 0.9$ does not guarantee the fit to a straight line to be very satisfactory.

7.3 General least-squares fit

When the function to be fitted $f(x, \boldsymbol{\theta})$ with parameters $\boldsymbol{\theta} = \theta_1, \ldots, \theta_n$ is more general than $ax + b$, the following cases should be distinguished:

(i) f **linear in $\boldsymbol{\theta}$.** For functions linear in all parameters, such as

$$f(x) = ax^2 + bx + c \qquad (7.28)$$

$$f(x) = a + b\exp(-k_1 x) + c\exp(-k_2 x); \qquad (7.29)$$

$$(k_1, k_2 \text{ known constants})$$

$$f(x) = ax + b/x + c, \qquad (7.30)$$

an analytical (weighted) least-squares solution minimizing S (7.8) is still possible, but requires some matrix algebra. Appendix A9 gives details.

(ii) f **linear in several variables.** Functions linear in more than one independent ("explanatory") variable, such as

$$f(\xi, \eta, \zeta) = a\xi + b\eta + c\zeta + d, \qquad (7.31)$$

where ξ, η, ζ are the independent variables and a, b, c, d are the parameters, similarly yield analytical least-squares solutions using some matrix algebra. Such functions are also treated in Appendix A9.

(iii) f **nonlinear but linearizable.** Functions that are nonlinear in the parameters can often be transformed to linear functions, as was done in Section 6.2 on page 73 in order to obtain linear graphs. For example, the function $f(t) = a\exp(-kt)$ is not linear in the parameter k. But if you take the logarithm:

$$\ln f(t) = -kt + \ln a, \qquad (7.32)$$

you obtain a function that *is* linear in k and has the form $ax + b$. So you can apply linear regression to the points $(t_i, \ln y_i)$ and determine k and $\ln a$. But take proper care of the weights: if all values of y have equal standard deviations σ, the values $\ln y_i$ have different weights:

$$\sigma_{\ln y} = \left| \frac{d \ln y}{dy} \right| \sigma_y = \sigma_y / y_i \qquad (7.33)$$

In this case you should take the weights w_i equal to (or proportional to) y_i^2. Negative values of y, which may occur by random deviations for large values of t, cannot be handled. It is not allowed to selectively omit negative values, as that will bias the result. The best way to proceed is to omit all points with larger values for t than the value for which a negative y first occurred. Even better is the use of a general nonlinear fitting procedure.

(iv) f **nonlinear: general case.** Functions that are nonlinear in the parameters cannot always be linearized. For example, the function

$$f(t) = a \exp(-kt) + b \qquad (7.34)$$

cannot be transformed to a linear function of the parameters a, b, and k. A nonlinear least-squares fitting procedure must then be used. In this case there are no analytical solutions and solutions are obtained by iterative function minimizers. There are several minimizers available, some that require analytical derivatives and some that do not. The latter are more easy to use. In all cases an initial guess of the parameters is required; for some functions a bad guess may lead to failure of the minimization procedure. A graphical analysis is a good source for a reasonable initial guess.

Below an example is given of a nonlinear least square minimization using Python.

Example: nonlinear fit

Consider the data on enzyme kinetics given in Table 6.2 on page 76. Given are six data points S_i, v_i with equal weights and our task is to fit two parameters $p_0 = v_{max}$ and $p_1 = K_m$ in the function

$$f(S, p) = \frac{p_0 S}{p_1 + S}; \qquad (7.35)$$

$$p = [v_{max}, K_m], \qquad (7.36)$$

such that the sum of squares

$$SSQ = \sum_i [v_i - f(S_i, p)]^2 \qquad (7.37)$$

is minimal. One possibility is to use the Python least-squares minimization procedure `leastsq` that comes with the module `optimize` of `SciPy`. This function requires a specification of the residues $y_i - f_i$ (or $(y_i - f_i)/\sigma_i$ if s.d.'s are known), which are a function of the parameters p, but does not require any derivatives. It must be called with an initial guess for p, for which we choose the values found by graphical inspection in Exercise 7.6:

$$p_{init} = [15, 105]. \qquad (7.38)$$

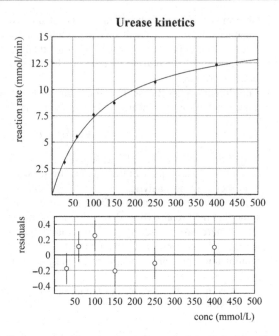

Figure 7.3 *Upper panel*: The urease reaction rate data plotted together with the least-squares fitted function. *Lower panel*: Plotting the residuals $y_i - f_i$ with error bars more clearly show whether the deviations have a random character or not.

The SSQ using this initial guess equals 0.375. After applying the minimization procedure the parameters appear to be

$$p_{\min} = [15.75, 114.65] \qquad (7.39)$$

and the minimal SSQ is 0.171. Figure 7.3 shows the fit, together with a plot of the residues. The latter plot is able to show the size of the error bars, and gives a visual impression of any systematic deviations. Later we shall see how large the uncertainties in the parameters appear to be.

Another possibility is to use the Python procedure `fmin_powell`, also in the module `optimize` of `SciPy`. With this minimizer the function to be minimized must be specified. This routine is less accurate than `leastsq` and should preferably be applied more than once.

A Python code which performs these minimizations is **Python code** 7.1 on page 183

Having determined the best values of the parameters, your problems are not yet solved! You also wish to assess the validity of the fit and you wish to have an estimate of the inaccuracies in the parameters. The key to the answers to these problems lies in the value of χ^2 as a function of the parameters. The next two sections explain all this.

7.4 The chi-squared test

Suppose you have performed a least-squares fit of the parameters in a function to a set of data. Is the fit reasonable, i.e., are the data – within experimental errors – compatible with the functional relation? What criteria can you apply to answer this question? What is "reasonable?"

Deviations from an acceptable fit can be expected both ways: the fit can be *not good enough* but it can also be *too good*. When the function has too few parameters or the functional shape is incorrect, the data will have systematic deviations that exceed the expected random errors. When there are too many parameters, the functional fit (if it succeeds at all) will follow the data too closely and the deviations will be less than expected from random errors.

You should always first check how the deviations $y_i - f_i$ (i.e., the residues) depend on x. A successful fit will yield residues that are samples from a random, generally normal, distribution. Systematic deviations are generally immediately apparent from a plot of the residues versus x. If you see such deviations, your fit is not acceptable and you should reconsider the functional choice you have made.

If there are no obvious deviations, the next step is to perform a *chi-squared test*. This test checks whether the sum of square deviations is compatible with the expectation based on the assumption that the deviations are samples from a (known) probability distribution. *The chi-squared test can only be performed if you have reliable prior estimates for the standard inaccuracies σ_i of each data point.* How to proceed when you don't have prior knowledge about the expected deviations is explained at the end of this section. Determine the minimal value of χ^2, defined as

$$\chi_0^2 \overset{\text{def}}{=} \sum_{i=1}^{n} \frac{(y_i - f_i)^2}{\sigma_i^2}, \tag{7.40}$$

where the f_i are evaluated at the optimal parameter values. This is just the value of S_0 (7.6) when all weights are taken equal to the inverse variances:

$$w_i = \sigma_i^{-2}. \tag{7.41}$$

Note: if you have determined S using weights that are proportional but not equal to inverse variances, you can derive χ_0^2 from

$$\chi_0^2 = \frac{S_0}{w} \sum_{i=1}^{n} \sigma_i^{-2}, \qquad (7.42)$$

where w is the total weight $\sum w_i$. See Exercise 7.6.

Since each of the terms in the chi-squared sum should have an expectation of 1, you expect the sum to be close to n. This is not quite correct: for a linear regression two degrees of freedom have been "used" for the determination of the two parameters a and b. The number of degrees of freedom ν equals $n - 2$ and χ^2 will be approximately equal to $\nu = n - 2$. In general, when there are m adjustable parameters, the remaining number of degrees of freedom is $n - m$. But χ^2 has a probability distribution around this value. This probability function depends on ν and is – if based on a normal distribution of the contributing deviations – denoted by $f(\chi^2|\nu)$. The larger the number of degrees of freedom, the narrower the chi-squared distribution becomes. The function $f(\chi^2|\nu)$ is fairly complex (see the data sheet CHI-SQUARED DISTRIBUTION on page 199 for equations), but has the convenient property that it approaches a normal distribution for large ν. The mean is equal to ν and the standard deviation equals $\sqrt{2\nu}$; this is not only true in the limit of large ν but is valid for all ν.

Tables of the chi-squared distribution do not give the probability density, but give the *cumulative distribution function* (cdf) $F(\chi^2|\nu)$: the probability that a sum of squares does *not* exceed χ^2. The survival function (sf) $1 - F(\chi^2|\nu)$ then indicates the probability that the value χ^2 *is* exceeded. The data sheet CHI-SQUARED DISTRIBUTION gives a table of values of χ^2 that are still acceptable for acceptance limits of 1%, 10%, 50%, 90% and 99%. Most books on statistics (and the *Handbook of Chemistry and Physics*) contain larger tables, but you may find it easier to use a Python routine from SciPy.

 See **Python code** 7.2 on page 184 to generate chi-squared probabilities.

Use the table or Python code as follows. First set an acceptance criterion, e.g. between 1% and 99%, or between 10% and 90%. The choice is subjective and you should report your choice when you publish your results. In the following example we choose the 10–90% limits. If your least-squares χ^2-value is less than the 10% value, you do not accept the outcome as a random occurrence and conclude that the fit is *too good*. Your function has too many parameters and you should try a simpler function with the same data: *your data do not justify the complexity of your function*. On the other hand, if your least-squares χ^2-value exceeds the 90% value, you do not accept the outcome as a random occurrence either and conclude that the data deviate significantly from the function. In that case: look for a function that better describes the data, possibly with more parameters. In both cases it is also good practice to review your data and your original estimates of the variances.

Example 1

In the urease kinetics example (see page 7.3) the least-squares fit gave a minimal value for the sum of squared deviations SSQ of 0.171. Since the s.d. of the data y_i were given as 0.2 mmol/min, the minimum value of χ^2 appears to be $0.171/0.2^2 = 4.275$. This is quite close to the number of degrees of freedom $\nu = n - m = 6 - 2 = 4$, so you may conclude that the deviations are compatible with random fluctuations. Indeed, the cdf of the chi-squared distribution $cdf(4.275,4)$ equals 0.63, a completely insignificant deviation.

Example 2

You have 10 independent measurements (x, y), with x being accurate and y having a known standard uncertainty σ. You have a simple theory that predicts y to be a linear function $ax + b$ of x, but a more refined theory predicts a second-degree function $px^2 + qx + r$. Do your data justify the refined theory above the simple one at a confidence level of 90%? You perform a linear least-squares fit to both functions, using $1/\sigma^2$ as weight factor for all points. For the linear function you find $S = \chi^2 = 14.2$ and for the quadratic function you find $S = \chi^2 = 7.3$. Inspection of the table in data sheet CHI-SQUARED DISTRIBUTION for 8 degrees of freedom shows that 14.2 lies above the 90% limit and hence is not acceptable according to the chosen acceptance criterion. The quadratic function (with 7 degrees of freedom) is indeed acceptable and the data justify its use. Had these values been 12.3 (linear function) and 6.5 (quadratic function), then the conclusion that a quadratic function should be used would not have been warranted, despite the fact that the quadratic function gives a better fit than the linear function.

What should you do if the standard uncertainties of the data are unknown or inaccurately known? In that case a chi-squared test cannot be used. It could well happen that you overestimate the experimental uncertainties by a factor of 2; this makes χ^2 a factor of 4 smaller. With e.g. 10 degrees of freedom you may then find a value of 2, while the accurate value is 8. The value 2 is below the 1% probability limit, making the fit unacceptable, while the accurate value of 8 is perfectly acceptable. Thus you may draw the wrong conclusion based on a wrong prior estimate of the uncertainties. This example shows that your prior knowledge of the inaccuracies in the data points should be rather precise for a valid chi-squared analysis.

With a sufficient number of data you may *use* the data themselves to determine the uncertainties of the measurements y. The minimal sum of squared deviations S_0 provides the information to estimate the individual variances $\hat{\sigma}_i^2$ if you take for χ^2 the best estimate $\hat{\chi}_0^2 = n - m$. The relation is

$$\hat{\sigma}_i^2 = \frac{S_0}{(n - m)w_i}, \tag{7.43}$$

which is easily obtained by setting $w_i = c/\sigma_i^2$ and solving c from

$$\hat{\chi}_0^2 = \frac{S_0}{c} = n - m. \tag{7.44}$$

Note that in the common case of equal weights and $w_i = 1$, $\hat{\sigma}^2 = S_0/(n-m)$.

Of course you cannot use $\hat{\chi}_0^2$ now to assess the quality of the fit. Therefore you should apply other criteria to analyze the random character of the residues $\varepsilon_i = y_i - f_i$. A graph versus x should not show systematic deviations. The cumulative distribution function should resemble a symmetric normal distribution. Statistical attributes such as mean and variance, if taken over sections of the data, should not differ significantly for different sections.

7.5 Accuracy of the parameters

Suppose you have performed a least-squares fit and your residues stood the tests for randomness so that you can trust the values of the standard uncertainties in the data. Either you had accurate prior knowledge of the standard uncertainties of the data points or you have scaled your uncertainties such that χ^2 in the minimum exactly equals $n - m$; in any case you know χ^2 as a function of the parameters. Now you can compute the variances and covariances of the fitted parameters.

In this section some general equations are given; their derivations are given in Appendix A9. We start again from n data points x_i, y_i and perform a least-squares fit of a (linear or nonlinear) function of m parameters $\theta_k, k = 1, \ldots, m$. The procedure yields χ^2 as a function of the parameters, with a minimum χ_0^2 for the parameter values $\hat{\theta}_i$, which are considered as best estimates. Because χ_0^2 is a minimum, the function $\chi^2(\theta_1, \theta_2, \ldots, \theta_m)$ is *quadratic* in the neighborhood of the minimum (the quadratic term is the first term in a Taylor expansion of χ^2 about the minimum, also for a fit function which is not linear in the parameters).

An important role in the derivation of (co)variances for the parameters is played by the matrix $\boldsymbol{B}$ with elements

$$B_{kl} = \sum_{i=1}^{n} \frac{1}{\sigma_i^2} \frac{\partial f_i}{\partial \theta_k} \frac{\partial f_i}{\partial \theta_l}. \tag{7.45}$$

The partial derivatives of f with respect to θ are constants for functions that are linear in the parameters; for nonlinear functions it is required to take the derivatives at the best-fit values $\hat{\theta}$ of the parameters. This matrix is also half the matrix of *second* derivatives of the function $\chi^2(\theta)$ (see Appendix A9):

$$B_{kl} = \frac{1}{2} \frac{\partial^2 \chi^2}{\partial \theta_k \partial \theta_l}, \tag{7.46}$$

meaning that B quantifies the curvature of the function $\chi^2(\theta_1, \theta_2, \ldots, \theta_m)$ at the minimum:

$$\Delta \chi^2 = \chi^2(\theta) - \chi^2(\hat{\theta}) \tag{7.47}$$

$$\approx \sum_{k,l=1}^{m} B_{kl} \Delta \theta_k \Delta \theta_l, \tag{7.48}$$

where $\Delta \theta_k = \theta_k - \hat{\theta}_k$. For linear functions the $\approx$ sign can be replaced by an $=$ sign.

Covariances of the parameters

The (co)variances of the parameters follow from the likelihood (7.4) on page 86:

$$p(\theta) \propto \exp \left[-\frac{1}{2} \Delta \chi^2(\theta) \right]. \tag{7.49}$$

By inserting (7.48) into (7.49) a bivariate normal distribution is obtained. As is more fully explained in Appendix A9, the (co)variances of the parameters are given by the *inverse* of the matrix B. Denoting this inverse by C:

$$C = B^{-1}, \tag{7.50}$$

then

$$\text{cov}(\theta_k, \theta_l) = C_{kl}. \tag{7.51}$$

From the covariance matrix the standard deviation σ_{θ_k} of θ_k (which is its standard inaccuracy) can be found:

$$\sigma_{\theta_k} = \sqrt{C_{kk}}. \tag{7.52}$$

The correlation coefficient ρ_{kl} between θ_k and θ_l is

$$\rho_{kl} = \frac{C_{kl}}{\sqrt{C_{kk} C_{ll}}}. \tag{7.53}$$

Table 7.1 *Relation between* $\Delta\chi^2$ *and the probability distribution of a single parameter* θ.

$\Delta\chi^2$	$p(\Delta\theta)/p(0)$	$\Delta\theta$	$P(-\Delta\theta, \Delta\theta)$
0.00	1.00000	0.0000	0.00%
0.50	0.77880	0.7071	52.05%
1.00	0.60653	1.0000	68.27%
1.50	0.47237	1.2247	77.93%
2.00	0.36788	1.4142	84.27%
2.50	0.28650	1.5811	88.62%
3.00	0.22313	1.7321	91.67%
3.50	0.17377	1.8708	93.86%
4.00	0.13534	2.0000	95.45%
4.50	0.10540	2.1213	96.61%
5.00	0.08208	2.2361	97.47%

Thus knowledge of the matrix C suffices to estimate inaccuracies of the parameters and their mutual correlations. In the following two paragraphs the one- and two-dimensional cases are graphically illustrated.

Relation between χ^2 and 1-D parameter distribution

In Table 7.1 the relations are given between $\Delta\chi^2$ and the distribution function of $\Delta\theta$ in the case of only one parameter, for which

$$p(\Delta\theta) = p(0)\exp\left[-\frac{1}{2}\Delta\chi^2(\Delta\theta)\right], \tag{7.54}$$

$$\Delta\chi^2(\Delta\theta) = b(\Delta\theta)^2. \tag{7.55}$$

Here b is the matrix element B_{11} in the expansion (7.48); $1/b$ is the variance σ_θ^2. The standard deviation is reached when $\Delta\chi^2 = 1$ (note that this is a fraction $1/\nu$ of $\Delta\chi_0^2$, the latter having an expectation of the number of degrees of freedom ν). Twice the standard deviation is reached for $\Delta\chi^2 = 4$. Figure 7.4 depicts the relevant relations.

This relation between $\Delta\chi^2$ and $p(\Delta\theta)$ is not only valid in the single-parameter case. When there are several parameters, and you wish to know the *marginal* probability distribution of a particular parameter θ_1, then it suffices to compute the function $\Delta\chi^2(\Delta\theta_1)$, while minimizing $\Delta\chi^2$ with respect to all other parameters. So the standard deviation of $\Delta\theta_1$ is reached

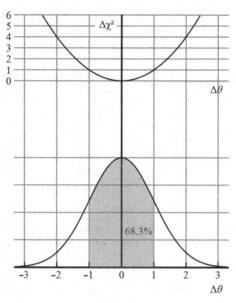

Figure 7.4 Relation between $\Delta\chi^2$ and the probability distribution of a single parameter θ.

when $\Delta\chi^2(\Delta\theta_1) = 1$, while all other parameters have their least-squares values. Why this is so is explained in Appendix A9.

Relation between χ^2 and 2-D parameter distribution

In the case of two parameters, $\Delta\chi^2$ is a quadratic function of two parameters $\Delta\theta_1$, $\Delta\theta_2$. The important feature to remember is that the vertical and horizontal *tangents to the contour* $\Delta\chi^2 = 1$ are positioned at $\theta_1 = \pm\sigma_1$ and $\theta_2 = \pm\sigma_2$, respectively. Appendix A9 explains why. Table 7.2 gives the relations between $\Delta\chi^2$, their tangent projections and the integrated percentage probability that the combined $(\Delta\theta_1, \Delta\theta_2)$ lies within the contour corresponding to $\Delta\chi^2$. The integrated probability $P(\Delta\chi^2)$ within the contour is a simple function of $\Delta\chi^2$:

$$P = 1 - \exp\left(-\frac{1}{2}\Delta\chi^2\right), \qquad (7.56)$$

with inverse:

$$\Delta\chi^2 = -2\ln(1 - P). \qquad (7.57)$$

For example, the θ values that represent 99 percent of the integrated joint probability are contained in the contour $\Delta\chi^2 = -2\ln 0.01 = 9.21$.

Table 7.2 *Relation between* $\Delta\chi^2$ *and the 2-D*
probability distribution of two parameters
$\Delta\theta_1, \Delta\theta_2$.

$\Delta\chi^2$ contour	*tangent projection* *in units* σ	*P(contour)* integrated
0.0	0.000	0.00%
0.5	0.707	22.12%
1.0	1.000	39.35%
1.5	1.225	52.76%
2.0	1.414	63.21%
2.5	1.581	71.35%
3.0	1.732	77.69%
3.5	1.871	82.62%
4.0	2.000	86.47%
4.5	2.121	89.46%
5.0	2.236	91.79%
5.5	2.345	93.61%
6.0	2.449	95.02%

An example is given in Fig. 7.5, with the following choice for B and
$C = B^{-1}$:

$$B = \frac{1}{3}\begin{pmatrix} 1 & -1 \\ -1 & 4 \end{pmatrix}; \quad C = \begin{pmatrix} 4 & 1 \\ 1 & 1 \end{pmatrix},$$

the latter meaning that

$$\sigma_1 = 2; \quad \sigma_2 = 1; \quad \rho_{12} = 0.5.$$

Figure 7.5 represents a *contour plot*: Each contour encloses all values of
the two parameters for which the *joint* probability exceeds a given level.
The integrated joint probability for the parameter values within the contour
$\Delta\chi^2 = 1$ (dark grey area) equals 39 percent (see Table 7.2); the projection
on the $\Delta\theta_1$-axis indicates the standard deviation $\sigma_1 = 2$. Thus the integrated
marginal probability of $\Delta\theta_1$ (light grey area) equals the usual 68 percent
between $\pm\sigma$ you already know from the normal distribution. It is also possible to read the value of the correlation coefficient ρ from the contour
$\Delta\chi^2 = 1$: The contour intersects the $\Delta\theta_1$-axis at the value $\sigma_1\sqrt{1 - \rho^2}$ and
likewise for $\Delta\theta_2$. The larger $|\rho|$, the more elongated is the ellipse in diagonal
direction. For positive ρ the long axis is in the SW–NE direction; for negative
ρ it is in the NW–SE direction.

$\Delta\chi^2$ contours

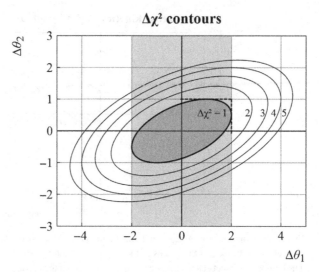

Figure 7.5 Contour plot of $\Delta\chi^2$ for the two-parameter case. The contour $\Delta\chi^2 = 1$ projects onto the axes at the values of the standard deviations ($\sigma_1 = 2$; $\sigma_2 = 1$).

 See **Python code** 7.3 on page 184 to generate a contour of a two-dimensional function at a prescribed level.

When there are more than two parameters, a 2-D contour plot with the same properties can still be made for any selected pair of parameters $\Delta\theta_1, \Delta\theta_2$, but it is then required that $\Delta\chi^2(\Delta\theta_1, \Delta\theta_2)$ is minimized with respect to all other parameters. The 2-D contours are projections of an m-dimensional ellipsoid representing $\Delta\chi^2$ in the space of all parameters.

Example

Consider the urease kinetics example (page 76; data from Table 6.2). A least-squares fit has already been performed (page 93). You have a function $S(v_{max}, K_m)$, which in this case is the unweighted sum of square deviations. The minimum of this function is $S_0 = 0.171$, occurring at the parameter values [15.75, 114.64]. You have also seen before (Example 1 on page 97) that S_0 is compatible with the known inaccuracies in the measurements, according to a chi-squared analysis. What you need is χ^2 as a function of the parameters and this you obtain by scaling S such that the minimum value scales to the expected value $n - m = 4$:

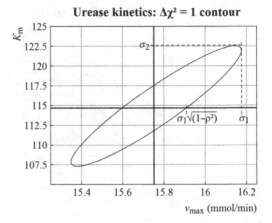

Figure 7.6 Contour plot of $\Delta\chi^2 = 1$ for the urease kinetics example. The contour projects onto the axes at the values of the standard deviations ($\sigma_1 = 0.41$; $\sigma_2 = 7.6$). The contour intersects each axis at a fraction $\sqrt{1 - \rho^2}$ of the corresponding standard deviation.

$$\chi^2(v_{max}, K_m) = \frac{n - m}{S_0}S = \frac{4}{0.171}S(v_{max}, K_m). \qquad (7.58)$$

Figure 7.6 shows the contour of $\chi^2 = 1$ in 2-D parameter space. You can read the standard uncertainties in the parameters from the projections of the ellipse on the axes and derive the correlation coefficient ρ from the intersection of the contour with the axes through the minimum. These values can also be taken from the array of contour points from which the contour plot is generated. The results are:

$$\sigma_1 = 0.41; \quad \sigma_2 = 7.63; \quad \rho = 0.93.$$

As you see, the two parameters are strongly correlated. A simultaneous deviation of the two parameters in the same direction (both positive or both negative) is much more probable than a simultaneous deviation in opposite directions. This is also clear from Fig. 7.6. The correlation coefficient is important if you need to predict the inaccuracy of the reaction rate at a given concentration.

See **Python code** 7.4 on page 186 to generate the χ^2-contour and to derive the uncertainties and correlation from the contour data.

If you had used your data errors (given as $\sigma_y = 0.2$ in this case) you would have found $\chi_0^2 = 0.171/0.2^2 = 4.3$ instead of the expectation 4.0.

Table 7.3 *Standard deviations and correlation coefficient for* v_{max} *and* K_m *in the urease kinetics example, computed with different methods.*

method	σ_1	σ_2	ρ
from leastsq routine	0.41	7.6	0.92
from $\Delta\chi^2 = 1$	0.41	7.6	0.92
from $C = B^{-1}$; $\delta = [0.2, 3.5]$	0.42	7.8	0.92
as above with $\delta = [0.0004, 0.007]$	0.36	7.0	0.93

This would have made your standard deviations an insignificant 4 percent larger.

In order to obtain the standard deviations and correlations without producing a $\Delta\chi^2 = 1$ contour, the covariance matrix C is required. This is the general procedure for more than two parameters when a 2-D plot is less suitable. If an appropriate least-squares program is used, the program can provide this matrix, as its elements are built up during the minimization procedure. Another route to the covariance matrix is the construction and subsequent inversion of the matrix $B = C^{-1}$, see (7.48). The elements of B can be found by evaluating $\Delta\chi^2$ at grid points near the minimum, e.g. at displacements δ_i in each of the parameters i and at displacements δ_i, δ_j for all pairs. When you perform the latter evaluation and you take the test displacements about equal to the standard deviations, you find results that are similar (but not equal) to covariances produced by the least-squares procedure (see Table 7.3). They are also close to the results obtained from the $\Delta\chi^2 = 1$ contour. However, if you take very small test displacements, the covariance matrix differs (in this example the values are 10 to 20 percent lower). The reason is that $\Delta\chi^2$ is not a pure quadratic function of the parameters as a result of the nonlinear character of the fit function. The likelihood $\exp\left(-\frac{1}{2}\Delta\chi^2\right)$ is not a pure bivariate normal distribution. It is best to use the covariances derived from $\Delta\chi^2$ in the neighborhood of 1, but be aware that the non-normal character will influence the tails of the likelihood distribution. Don't trust confidence intervals based on normal distributions and take the standard deviations with a pinch of salt.

See **Python code** 7.5 on page 187 to generate the covariance matrix for the urease kinetics example from the least-squares routine.

See **Python code** 7.6 on page 187 to generate the covariance matrix for the urease kinetics example from construction of the B matrix.

See **Python code** 7.7 on page 189 for a program that reports the results of a general least-squares fit of a predefined function to given data.

7.6 F-test on significance of the fit

If you have fitted a theoretical relation to your data points, the first question to ask is whether the fit has any significance at all. Does the fit significantly reduce the sum of square deviations compared to the sum of square deviations with respect to the average of y_i? If not, the function does not add anything to explain the data. But what is "significant" in this context?

The *total* sum of squared deviations of the data y_i with respect to their average $\langle y \rangle$ is given by (for simplicity we take all weight factors equal to one):

$$\text{SST} = \sum_{i=1}^{n} (y_i - \langle y \rangle)^2. \tag{7.59}$$

This is the relevant sum of square deviations if you had no model at all. The number of degrees of freedom is $n - 1$ since you have used the data to determine one parameter (the average). *With* a model you predict values f_i that are matched as closely as possible to y_i. The residual sum of squared deviations, now called the *error sum of squares* SSE, is

$$\text{SSE} = \sum_{i=1}^{n} (y_i - f_i)^2. \tag{7.60}$$

If your functional relation contains m parameters to determine the f_i's, the number of degrees of freedom is $n - m$. This sum of squares is the least you can obtain with your model; it is entirely due to random errors. The difference $\text{SST} - \text{SSE}$ is the part of the total sum of squared deviations that is explained by the model. It is called SSR, the *regression sum of squares*. Its magnitude is

$$\text{SSR} = \text{SST} - \text{SSE} = \sum_{i=1}^{n} (f_i - \langle f \rangle)^2. \tag{7.61}$$

The number of degrees of freedom associated with SSR is $m - 1$ since the f_i's are determined by m variables, but one is used for the average. You see that all degrees of freedom are accounted for.

The validity of this equality is not immediately clear. It is valid when the f_i's have been determined such that the residues $\varepsilon_i = y_i - f_i$ form a set of independent samples from a probability distribution with zero mean. The latter implies that $\langle y \rangle = \langle f \rangle$, while independence means that ε_i is not related to f_i: $\sum_i \varepsilon_i (f_i - \langle f \rangle) = 0$. From this it follows that

$$\text{SST} = \sum_i (y_i - \langle y \rangle)^2 = \sum_i (y_i - f_i + f_i - \langle f \rangle)^2$$

$$= \text{SSE} + \text{SSR} + 2\sum_i (y_i - f_i)(f_i - \langle f \rangle) \qquad (7.62)$$

$$= \text{SSE} + \text{SSR}. \qquad (7.63)$$

In practice this relation may not be exactly fulfilled because the last term in (7.62) may not be exactly zero.

Having separated the total sum of squared deviations into a part SSR explained by the model and a random-error part SSE, it is clear that the larger SSR is (relative to SSE), the more significant your model is. The proper statistical test is the F test (see Chapter 4, page 106, and the datasheet F-DISTRIBUTION on page 201). The F-ratio (which is the ratio of estimated variances) is given by

$$F_{m-1,n-m} = \frac{\text{SSR}/(m-1)}{\text{SSE}/(n-m)}. \qquad (7.64)$$

The cumulative F-distribution gives the probability that both sets of deviations are sampled from distributions with the same variance. It tests the null hypothesis "the model does not explain the data to a significant extent" or – equivalently – "the model is insignificant." You may reject the null hypothesis and accept the model as significant when the F-ratio exceeds the critical value F_c for which $F(F_c) > 1 - \alpha$, where α is the significance level. For example, if you have 10 data points and 3 adjustable parameters, so that $\nu_{\text{SSR}} = 2$ and $\nu_{\text{SSE}} = 7$, and you set the significance level at 1 percent, then the critical F-ratio equals 9.55 (see the second table of data sheet F-DISTRIBUTION on page 201). For larger values than 9.55 you can be confident that the model is significant.

Example

Let us return to the urease kinetics example (page 76; data from Table 6.2). The fact that the two parameters $v_{max} = 15.8 \pm 0.4$ and $K_m = 115 \pm 8$ are highly significant (see example on page 103) already suggests that the fit is very relevant. Indeed, almost all variation of the measured values is explained by the model, as you can see by evaluating the sum of squared deviations: SST = 57.02; SSR = 56.24; SSE = 0.17. Note that the sum of SSR and SSE is not quite equal to SST, which is to be expected for a nonlinear least-squares fit. The F-ratio equals [SSR/1]/[SSE/4] = 1315 and the cumulative probability of the F-distribution is 0.9999966. You can be (very) confident that the model is relevant! See Exercise 7.7 for an example with a more dubious outcome.

See **Python code** 7.8 on page 193 to compute these results.

Summary *You are now able to perform least-squares fits of parameters in a functional relation to a given data set. For functions that are linear in the parameters, the least-squares method is robust as long as the parameters are not mutually dependent. For nonlinear functions a minimum can usually be found by appropriate iterative computer programs. You can test your fit in two ways: the first question is whether the fit to the specified function is significantly better than a fit to just an average: use an F-test for this assessment. The second question is whether the residual deviations with respect to the fit behave as random samples from a distribution with variance compatible with your prior knowledge of the uncertainties: use a chi-squared test for this assessment. If OK, then you can compute the covariance matrix of the parameters from the observed sum S of square deviations. Use the dependence of S on the parameters to find this matrix.*

Exercises

7.1 Perform a linear regression on the enzyme kinetics data of Table 6.2 (page 76). Do this according to the Lineweaver–Burk plot, i.e., for $x = 1/[S]$ en $y = 1/v$. Use correct standard inaccuracies for y. Give the values for v_{max} and K_m with their standard inaccuracies (be aware that for an s.d. in a combination of a and b also the correlation coefficient between a and b is needed). Compare your values with the graphical estimates from Fig. 6.2 and with the nonlinear least-squares solution of the example on page 93. Plot the data points with the best-fitted straight line.

7.2 The ΔG of a reaction was determined by measuring equilibrium constants at various temperatures. The following values were obtained:

T/K	$\Delta G/kJ\ mol^{-1}$
270	40.3
280	38.2
290	36.1
300	32.2
310	29.1
320	28.0
330	25.3

The uncertainty in T is negligible and the weight factors are equal for all cases. Determine the reaction entropy $\Delta S = -d\Delta G/dT$ by fitting the values of ΔG to a linear function of T. What is the standard inaccuracy in ΔS? Extrapolate ΔG to $T = 350\,\text{K}$ and give the standard inaccuracy that follows from the variance and covariance of the parameters, as found from the least-squares fit. Now do the same thing, but take for x the values of $T - 300$ instead of T itself. Discuss the differences (if any) between the two calculations.

7.3 Explain why the weight given to a data point $y_i = \log t_i$ should be proportional to t_i^2 if the t_i's are all random samples from probability distributions with the same variance. Start by deriving the variance of y assuming a constant variance σ_t^2 of t. Then relate the weight to the variance.

7.4 Perform a least-squares fit of the four-parameter double exponential function $a\exp(-px) + b\exp(-qx)$ to the data x, y from Table 6.1 on page 74. Use the Python program $\texttt{fit}$ (code **7.7**). As initial parameter guess use the values that were determined graphically in Section 6.2. If the minimization does not reach a result after the maximum number of trials, then take the last values of the parameters as initial ones and minimize again.

7.5 You wish to measure the focal length of a positive lens on an optical bench with a ruler graduated in mm from 0 to 1000. Your lens is placed somewhere near 190 mm but it is an encased thick lens and you are not sure about its exact position. Your object (a lamp) is placed at position x and you observe the image to be sharp at position y. You estimate the s.d. σ_y of y. All data are in mm.

x	y	σ_y
60	285	1
80	301	2
100	334	3
110	383	4
120	490	5
125	680	10

Set up the parameterized equation $y \approx f(x, p)$, assuming the thin-lens formula to be valid: $1/f = 1/s_1 + 1/s_2$, where s_1 and s_2 are the distances from the lens to the object and the image, respectively. Find the least-squares solution and evaluate the best value for f and its standard inaccuracy. Discuss the validity of the functional fit. Use Python.

7.6 Prove (7.42) by setting $w_i = c/\sigma_i^2$ and eliminating c.

7.7 The F-test for relevance of a least-squares fit can be used to test if there is a *drift* in a time series, e.g. a time-dependent variable generated by a simulation that is supposed to produce stationary fluctuating quantities. Generate 100 random numbers from a normal distribution $N(0, 1)$. Perform a least-squares fit to $f_i = ai + b$, yielding the best estimates $\hat{a}$ and $\hat{b}$, and compute the standard inaccuracy in a. There are two ways to check if the drift you find is significant. First you can assess the probability that a differs from zero, i.e., the two-sided probability that $|a| \leq |\hat{a}|$. Although Student's t-distribution is the appropriate distribution, you can also use a normal distribution because with a large number of degrees of freedom the t-distribution is nearly equivalent to a normal distribution. The second method is using an F-test. Compute SSR and SSE and assess the significance of the linear regression model with an F-test. Do all this for a few fresh series of random samples and compare the two results. In order to get significant results you may add a drift term to the data.

Remark: This is done most easily by calling the routine `report`, see **Python code 6.2** on page 176.

8 Back to Bayes: knowledge as a probability distribution

In this chapter the reader is requested to sit back and think. Think about what you are doing and why, and what your conclusions really mean. You have a theory, containing a number of unknown – or insufficiently known – parameters, and you have a set of experimental data. You wish to use the data to validate your theory and to determine or refine the parameters in your theory. Your data contain inaccuracies and whatever you infer from your data contains inaccuracies as well. While the probability distribution of the data, given the theory, is often known or derivable from counting events, the *inverse*, i.e., the inferred probability distribution of the estimated parameters given the experimental outcome, is of a different, more subjective kind. Scientists who reject any subjective measures must restrict themselves to hypothesis testing. If you want more, turn to Bayes.

8.1 Direct and inverse probabilities

Consider the reading of a sensitive digital voltmeter sensing a constant small voltage – say in the microvolt range – during a given time, say 1 millisecond. Repeat the experiment many times. Since the voltmeter itself adds a random noise due to the thermal fluctuations in its input circuit, your observations y_i will be samples from a probability distribution $f(y_i - \theta)$, where θ is the real voltage of the source. You can determine f by collecting many samples. In some cases, when you know the physical process that adds the noise, you may even be able to predict the distribution function. For example, if you observe the number of light pulses in a given time interval Δt, knowing that they occur randomly at a given average rate θ, then the number k observed in a given time interval will obey the Poisson probability distribution $f(k, \theta \Delta t)$. Such (conditional) probabilities $f(y|\theta)$ are called *direct* probabilities. They result from direct counting of events or from considering symmetries in the random process. In this chapter the notation f is used for such direct probabilities which are also called *physical* probabilities.

Now consider the value of a physical constant, e.g. Avogadro's number N_A. It is the number of atoms in 1 gram of pure ^{12}C. According to CODATA its value is $(6.022\,141\,79 \pm 0.000\,000\,30) \times 10^{23}$. That is, the number given is not exact. At best a probability distribution $p(N_A)$ can be given for N_A, e.g. a normal distribution with mean $6.022\,141\,79 \times 10^{23}$ and standard deviation 3.0×10^{16}. But what does a probability of this kind mean? It is *not* a frequency distribution that you can find by counting the outcome of a large number of similar experiments, because – if there had been a large number of independent evaluations – the CODATA committee would have averaged those and proposed another mean and s.d. Similarly, the metereologist's prediction that "there is a 30 percent probability that it will rain tomorrow" or the surgeon's prediction that "the patient has a 95 percent chance of surviving the operation" says something about a unique event that cannot be repeated; such probabilities are more expressions of a belief based on earlier experience than the outcome of counting numbers in repeated experiments. Philosophers call such probabilities *epistemic*.[1] Other names are *subjective* and *inverse* probabilities.

The distinction between direct and inverse probabilities has been clear to Laplace and his followers since the late eighteenth century.[2] But the subjective nature of inverse probability has caused many scientists to shy away from using such concepts. The exponent of the critical school is the eminent statistician R. A. Fisher who developed a range of statistical tools in the first half of the twentieth century, all based on the frequency definition of direct probabilities. He circumvented the use of inverse probabilities by introducing the *likelihood* as a substitute.

There is a good reason to be critical to concepts in physics that are not entirely objective: subjective bias, arbitrariness and prejudice may easily creep into the interpretation of results. So, *if* you use inverse probabilities as an expression of your knowledge, it is essential that such probabilities are unbiased and do not include "information" that does not rest on verifiable knowledge. But with this restriction the use of inverse probabilities is very rich and very powerful to infer model parameters from experimental data.

Since the middle of the twentieth century the construction of inverse probabilities has gained ground over the critical school and is recently enjoying a real revival. It is called *the Bayesian approach*.

8.2 Enter Bayes

In two papers of Thomas Bayes (1763, 1764), published posthumously by R. Price, the principle of what is now called "Bayes' method" of constructing inverse probabilities was laid down within a context of combinatorial

---

[1] The word epistemic, from the Greek *epistème*: knowledge, was first used in the context of probabilities by Skyrms (1966).

[2] See Hald (2007) for a historical review of statistical inference.

problems. Ten years later the concept was worked out by Laplace. It is really very simple, once you agree to work with inverse probabilities.[3]

Consider two events, T and E (T stands for "theory": a parameter or set of parameters in a theory, and E for "experiment": an observed quantity or quantities). The probability of the joint event $p(T, E)$ can be expressed as the marginal probability of one event times the conditional probability of the other:

$$p(T, E) = p(T) p(E|T) = p(E)p(T|E). \qquad (8.1)$$

This implies that the *posterior* probability of T, $p(T|E)$, i.e., the probability after the experiment is known, is proportional to the product of the *prior* probability of T, $p(T)$, i.e., the probability before the experiment is known, and the probability of the experimental outcome, given the theory:

$$p(T|E) \propto p(T)p(E|T). \qquad (8.2)$$

The proportionality constant is really a normalization factor; it is simply the inverse of the sum or integral of the right-hand side over all possibilities of T.

In more specific terminology (and now the notation f is used for direct probabilities and p for inverse probabilities): you have a theory with a set of parameters θ and you have a set of data y. Now the posterior probability of your parameters is

$$p(\theta|y) = cf(y|\theta)p_0(\theta), \qquad (8.3)$$

where $p_0(\theta)$ is the *prior* probability density function of your parameters. The latter expresses the knowledge you have about θ before you know the experimental results. The constant c is given by

$$c^{-1} = \int f(y|\theta)p_0(\theta)\, d\theta, \qquad (8.4)$$

with integration carried out over the full domain of possible values for θ. Here it is assumed that the parameters can take on a continuous range of values (with p being probability densities), but they can just as well be discrete, in which case the integration becomes a summation and the p's are probability mass functions. Likewise the direct probabilities $f(y|\theta)$ can be either continuous or discrete.

---

[3] See, among many others, Box and Tiao (1973) and Lee (1989) for Bayesian treatments of statistical problems. Cox (2006) compares the frequentist and Bayesian approaches to statistical inference.

8.3 Choosing the prior

The prior distribution p_0 *must* be unbiased. It can only depend on previous experiments and derived by equations like (8.3). If no such experimental information is known, the prior must be as *uninformative* as possible: any information you put into the prior that does not rely on verifiable data introduces a form of prejudice.

The most uninformative prior is a constant: all values are equally possible. It seems a bit strange to propose a constant for a probability density function (pdf): a respectable pdf should be normalized, i.e., the integral over its domain should be unity. Probability densities that cannot be normalized are called *improper*. But you can make the constant pdf respectable if you cut its value to zero at the far ends beyond the range of possible values. Since the direct probability $f(y|\theta)$ is a peaked function with finite integral, the integral in (8.4) exists even for $p_0 \equiv 1$. So it is OK to allow improper priors.

There is one objective requirement for an acceptable prior: it should scale properly with transformations of the parameters. Consider a *location parameter* μ, occurring as an additive factor in the range $(-\infty, \infty)$. It could just as well be replaced by a linear transformation $\mu' = a\mu + b$; a uniform distribution of μ should also be uniform if expressed in μ'. That is indeed the case: since $p(\mu)\,d\mu = p'(\mu')\,d\mu'$ and $d\mu' = a\,d\mu$, $p' = p/a$, which is uniform if p is a constant. Now consider a *scale parameter* σ, occurring as a multiplicative factor in the range $(0, \infty)$. It could just as well be replaced by $c\sigma$ or by σ^2 or by σ^{-1}, or by the transformation $\sigma' = b\sigma^a$. It is clear that the variable $\log \sigma$ transforms linearly: $\log \sigma' = a \log \sigma + b$; therefore the distribution should be uniform in $\log \sigma$. This implies that the uncommitted (or *ignorant*) prior should be proportional to $1/\sigma$ since $d \log \sigma = d\sigma/\sigma$. Summarizing (this rule is due to Jeffreys, 1939):

The most uninformative (ignorant, uncommitted, unbiased) improper prior $p_0(\theta)$ equals 1 if θ is a location parameter or $1/\theta$ if θ is a scale parameter.

8.4 Three examples of Bayesian inference

Updating knowledge: Avogadro's number

CODATA suggests that we may believe the inverse probability density function of Avogadro's number to be

$$p_0(N_A) \propto \exp\left[-\frac{(N_A - \mu_0)^2}{2\sigma_0^2} \right], \tag{8.5}$$

with $\mu_0 = 6.022\,141\,79 \times 10^{23}$ and $\sigma_0 = 3.0 \times 10^{16}$.

A scientist comes along with a reliable new measurement of N_A. She measured the value $y = 6.022\,141\,48 \times 10^{23}$ and asserts that her analysis of experimental errors indicates that her result y is a sample from a normal distribution $N(y - N_A, \sigma_1)$, where $\sigma_1 = 7.5 \times 10^{16}$.

Inserting these data into (8.3) we find that

$$p(N_A|y) \propto \exp\left[-\frac{(y - N_A)^2}{2\sigma_1^2} \right] \exp\left[-\frac{(N_A - \mu_0)^2}{2\sigma_0^2} \right]. \qquad (8.6)$$

Working out the exponent $\left(\text{omitting a factor } -\frac{1}{2} \text{ for the time being}\right)$:

$$\frac{(y - N_A)^2}{\sigma_1^2} + \frac{(N_A - \mu_0)^2}{\sigma_0^2} \qquad (8.7)$$

$$= \left(\sigma_0^{-2} + \sigma_1^{-2}\right)\left[N_A^2 - 2N_A \frac{\mu_0\sigma_0^{-2} + y\sigma_1^{-2}}{\sigma_0^{-2} + \sigma_1^{-2}} + \cdots \right] \qquad (8.8)$$

$$= \frac{(N_A - \mu)^2}{\sigma^2} + \cdots, \qquad (8.9)$$

where

$$\mu = \frac{\mu_0\sigma_0^{-2} + y\sigma_1^{-2}}{\sigma_0^{-2} + \sigma_1^{-2}}, \qquad (8.10)$$

$$\sigma^{-2} = \sigma_0^{-2} + \sigma_1^{-2}. \qquad (8.11)$$

Thus the posteriori inverse probability density of N_A is a normal distribution with weighted averages for the mean and variance (see also Exercise 5.7 on page 70):

$$p(N_A|y) \propto \exp\left[-\frac{(N_A - \mu)^2}{2\sigma^2} \right]. \qquad (8.12)$$

The result is that the parameters μ_0 and σ_0 in the prior pdf (8.5) have been updated to μ and σ in the posterior pdf (8.12).

Inference from a series of normally distributed samples

Suppose your experimental data are n independent samples from a normal distribution with unknown μ and unknown σ. You have no prior knowledge of the mean and s.d., so you take the uninformative prior

$$p_0(\mu, \sigma) = 1/\sigma, \qquad (8.13)$$

because μ is a location parameter and σ is a scale parameter. The probability of observing n values $y_i, i = 1, \ldots, n$ is the product of the probabilities of all measurements, because the data are independent:

$$f(y|\mu, \sigma) = \Pi_{i=1}^{n} \frac{1}{\sigma\sqrt{2\pi}} \exp\left[-\frac{(y_i - \mu)^2}{2\sigma^2}\right] \tag{8.14}$$

$$\propto \sigma^{-n} \exp\left[-\frac{1}{2\sigma^2} \sum_{i=1}^{n}(y_i - \mu)^2\right] \tag{8.15}$$

This can be rewritten as

$$\sigma^{-n} \exp\left[-\frac{(\langle y \rangle - \mu)^2 + \langle(\Delta y)^2\rangle}{2\sigma^2/n}\right], \tag{8.16}$$

where

$$\langle y \rangle = \frac{1}{n} \sum_{i=1}^{n} y_i \tag{8.17}$$

and

$$\langle(\Delta y)^2\rangle = \frac{1}{n} \sum_{i=1}^{n}(y_i - \langle y \rangle)^2. \tag{8.18}$$

For the posterior probability density we can now write:

$$p(\mu, \sigma | y) \propto \sigma^{-(n+1)} \exp\left[-\frac{(\mu - \langle y \rangle)^2 + \langle(\Delta y)^2\rangle}{2\sigma^2/n}\right]. \tag{8.19}$$

The proportionality constant can be obtained by integrating the right-hand side over both μ and σ. In this case there is an analytical expression for the integral, but it is often easier to determine the constant by numerical integration.

It is interesting to see that the probability (8.19) of the parameters is given by only two properties of the data set: the average and the mean-squared deviation from the average. Apparently these two properties are sufficient to know everything about the statistics of the data set (*sufficient statistics*). But this is only true if we already know that the samples come from a normal distribution!

The pdf of (8.19) is *bivariate*. It is plotted as a number of contours at various fractional heights in Fig. 8.1 for the example of 10 samples with $\langle y \rangle = 0$ and $\langle(\Delta y)^2\rangle = 1$. The values within a given contour represent a defined integrated probability.

In practice you will more often find use for one-dimensional distribution functions. First consider the pdf for μ (Fig. 8.2).

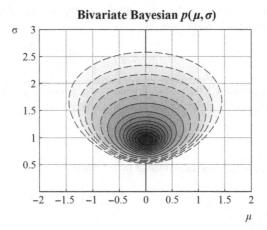

Figure 8.1 Contour plot of the Bayesian inverse bivariate pdf of the mean μ and s.d. σ given the value of 10 independent normally distributed experimental samples. The average equals zero and the rmsd equals 1. Contours – from inside out – are full-drawn at fractional heights 0.9, 0.8, ..., 0.1; broken at 0.05, 0.02, 0.01, 0.005, 0.002.

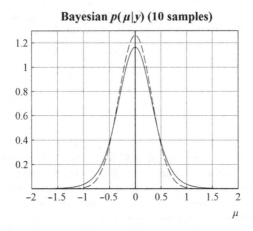

Figure 8.2 The Bayesian posterior pdf for the parameter μ, given the value of 10 independent normally distributed experimental samples with zero average and rmsd $= 1$. The drawn line is the marginal $p(\mu|y)$ for unknown σ; the broken line is $p(\mu|y,\sigma)$ for known $\sigma = 1$.

For *known* σ you see from (8.19) that the posterior pdf of μ is a normal distribution around $\langle y \rangle$ with variance σ^2/n:

$$p(\mu|\mathbf{y},\sigma) \propto \exp\left[-\frac{(\mu - \langle y \rangle)^2}{2\sigma^2/n}\right]. \tag{8.20}$$

For *unknown* σ the probability must be integrated over all possible values of σ in order to obtain a *marginal* distribution of μ.

$$p(\mu|\mathbf{y}) = \int_0^\infty p(\mu,\sigma|\mathbf{y})\,d\sigma. \tag{8.21}$$

This integral can be written as proportional to

$$\int_0^\infty \sigma^{-(n+1)} \exp\left(-\frac{q}{\sigma^2}\right)\,d\sigma, \tag{8.22}$$

where

$$q = \frac{1}{2}n[(\mu - \langle y \rangle)^2 + \langle(\Delta y)^2\rangle]. \tag{8.23}$$

By substituting q/σ^2 for a new variable, a Gamma-function is obtained. The integral appears to be proportional to

$$p(\mu|\mathbf{y}) \propto \left(1 + \frac{(\mu - \langle y \rangle)^2}{\langle(\Delta y)^2\rangle}\right)^{-n/2}. \tag{8.24}$$

This is exactly Student's t-distribution density function $f(t|v)$ for $v = n - 1$ degrees of freedom, as a function of the variable t:

$$f(t|v) \propto \left(1 + \frac{t^2}{v}\right)^{-(v+1)/2} \tag{8.25}$$

$$t = \sqrt{\frac{(n - 1)(\mu - \langle y \rangle)^2}{\langle(\Delta y)^2\rangle}} = \frac{\mu - \langle y \rangle}{\hat{\sigma}/\sqrt{n}}, \tag{8.26}$$

where $\hat{\sigma}^2 = [n/(n - 1)]\langle(\Delta y)^2\rangle$. See data sheet STUDENT'S T-DISTRIBU-TION on page 213 for further details on the t-distribution.

Next consider the pdf for σ. If μ is known, the pdf is given by (8.19). Figure 8.3 shows $p(\sigma|\mathbf{y}, \mu = 0)$ for the example used above. It is more common that you do *not* know μ in advance; then your Bayesian posterior probability is the marginal probability:

$$p(\sigma|\mathbf{y}) = \int_{-\infty}^\infty p(\mu,\sigma|\mathbf{y})\,d\mu \tag{8.27}$$

$$\propto \sigma^{-n} \exp\left[-\frac{\langle(\Delta x)^2\rangle}{2\sigma^2/n}\right]. \tag{8.28}$$

Bayesian $p(\sigma|y)$ (10 samples)

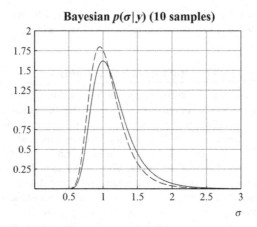

Figure 8.3 The Bayesian posterior pdf for the parameter σ, given the value of 10 independent normally distributed experimental samples with zero average and rmsd $= 1$. The drawn line is the marginal $p(\sigma|y)$ for unknown μ; the broken line is $p(\sigma|y, \mu)$ for known $\mu = 0$.

Bayesian pdf for rate process

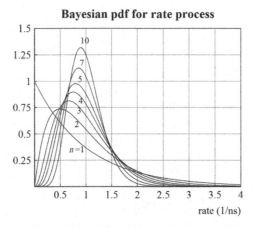

Figure 8.4 The Bayesian posterior pdf for the rate parameter k, given the value of n independent time intervals between events. For this example the average observed time equals 1 ns. The pdf's are drawn for $n = 1, 2, 3, 4, 5, 7, 10$.

As you see from Fig. 8.3, the value predicted for σ is slightly larger and slightly less accurate when μ is not known *a priori*.

Infer a rate constant from a few events

Consider the observation of single events that sample a rate process. This could be a pulse emitted from a source that is excited at $t = 0$; it could be the observation of a conformational change in the simulation of a protein that is made unstable by changing its environment at $t = 0$; it could be the time between two sightings of a meteor, or any other seldom event that you can observe only a few times. Your theory says that the event results from a simple rate process with constant probability $k\Delta t$ that the event occurs in any small time interval Δt. You observe n independent events at times or intervals $t_i, i = 1, \ldots, n$. What can you say about the rate constant k?

In a Bayesian approach you wish to determine after the first event the inverse posterior probability

$$p_1(k|t_1) \propto f(t_1|k)p_0(k), \tag{8.29}$$

where $f(t|k)$ is the direct probability that an event occurs after a time t, given the rate constant k. This is easy to derive. Divide time in small intervals Δt; $t/\Delta t = m$. The probability that the pulse occurs at the m-th interval, and not before, is $(1 - k\Delta t)^{m-1} k\Delta t$. Taking the limit $\Delta t \to 0; m \to \infty$, you find

$$f(t|k) = ke^{-kt}. \tag{8.30}$$

The prior inverse probability $p_0(k)$ must be taken as $1/k$, since k is a scale parameter. So

$$p_1(k|t_1) \propto e^{-kt_1}. \tag{8.31}$$

After observing a second event at time t_2, you can update this probability:

$$p_2(k|t_1, t_2) \propto ke^{-kt_2}e^{-kt_1} \tag{8.32}$$

and after n events you obtain

$$p_n(k|t_1, \ldots, t_n) \propto k^{(n-1)} \exp[-k(t_1 + \cdots, t_n)]. \tag{8.33}$$

In general, if the average of the observed time intervals is $\langle t \rangle$, and the proportionality constant is included by integrating this function, it is found that

$$p_n(k|t_1, \ldots, t_n) = \frac{(n\langle t \rangle)^n}{(n - 1)!} k^{n-1} \exp(-kn\langle t \rangle). \tag{8.34}$$

So you see that the average of the observation times is sufficient statistics: it determines all you can know about k. It is easily seen that the expectations over this distribution of k and its variance are given by

$$\hat{k} = E[k] = \frac{1}{\langle t \rangle}, \qquad (8.35)$$

$$\hat{\sigma}^2 = E[(k - \langle k \rangle)^2] = \frac{1}{n \langle t \rangle^2}. \qquad (8.36)$$

The latter equation implies that $\hat{\sigma} = \hat{k}/\sqrt{n}$. As always, the relative standard inaccuracy decreases with the square root of the number of observations.

The case $n = 7$ has been used before in this book: Fig. 2.5 on page 12. For that case three different *point estimates* are given: the mean (1.00), the median (0.95) and the mode (0.86). It is pointless to haggle about what is best, as all values are well within a s.d. (0.38) from the mean.

8.5 Conclusion

The examples above express your knowledge in terms of probability density functions. These have one disadvantage: they look much more exact than they are. Be aware that such probability distributions only express your degree of ignorance about the parameters derived from theory and experiment. Your best value is not necessarily the exact mean or the exact mode; it can be anywhere within the width of the distribution. Be careful to report the right number of digits!

How to proceed if you absolutely refuse to use inverse probabilities? First, you can fool yourself by defining the *likelihood* of a parameter as equal to the direct probability of the measured value:

$$l(\theta|y) = f(y|\theta). \qquad (8.37)$$

This is of course equivalent to the posterior Bayesian probability when you assume a uniform prior. Inconsistencies are expected for scale parameters. Renaming a quantity does not solve your problem but hides it like an ostrich does.

Second, you can limit yourself to testing hypotheses rather than predicting values. It is useful if you wish to assess the effect of some agent that does or does not influence a sampled result. The *null hypothesis* usually assumes that the agent has no effect and you try to prove that the result you obtain is unlikely under the null hypothesis. If so, you accept the truth of the alternative hypothesis ("the agent does have an effect"). This procedure avoids any inverse probabilities, but it makes life quite poor: you also want to know *what* the effect of the agent is. Many questions you wish to be answered by your experiments remain out of bounds.

Summary *This chapter has taken a Bayesian point of view on statistics, accepting the notion of "inverse probability". Simple rules then allow you to express all the knowledge you have, including the outcome of your recent experiments, in probability functions of the parameters in your theory. In three examples it is shown that you can either update existing knowledge with new experimental data or – without prior knowledge – express the knowledge gained from limited experimental data in probability distributions. The introduction to this chapter invited you to sit back and think. Now, sit back and draw your conclusions.*

References

Abramowitz, M. and Stegun, I. A. (1964). *Handbook of Mathematical Functions.* New York, Dover Publications.

Barlow, R. (1989). *Statistics – A Guide to the Use of Statistical Methods in the Physical Sciences.* New York, Wiley.

Bayes, T. (1763). *Phil. Trans. Roy. Soc.* **53**, 370–418. Reprinted in *Biometrika* **45**, 293–315 (1958).

Bayes, T. (1764). *Phil. Trans. Roy. Soc.* **54**, 296–325.

Berendsen, H. J. C. (1997). *Goed Meten met Fouten.* University of Groningen.

Berendsen, H. J. C. (2007). *Simulating the Physical World.* Cambridge, Cambridge University Press.

Bevington, P. R. and Robinson, D. K. (2003). *Data Reduction and Error Analysis for the Physical Sciences*, 3rd edn. (first edn. 1969). New York, McGraw-Hill.

Beyer, W. H. (1991). *CRC Standard Probability and Statistics Tables and Formulae.* Boca Raton, Fla., CRC Press.

Birkes, D. and Dodge, Y. (1993). *Alternative Methods of Regression.* New York, Wiley.

Box, G. E. P. and Tiao, G. C. (1973). *Bayesian Inference in Statistical Analysis.* Reading, Mass., Addison-Wesley.

Cox, D. R. (2006). *Principles of Statistical Inference.* Cambridge, Cambridge University Press.

Cramér, H. (1946). *Mathematical Methods of Statistics.* Princeton, NJ, Princeton University Press.

CRC Handbook (each year). *Handbook of Chemistry and Physics.* Boca Raton, Fla., CRC Press.

Efron, B. and Tibshirani, R. J. (1993). *An Introduction to the Bootstrap.* London, Chapman & Hall.

Frenkel, D. and Smit, B. (2002). *Understanding Molecular Simulation. From Algorithms to Applications.* 2nd edn., San Diego, Academic Press.

Gardner, M. (1957). *Fads and Fallacies in the Name of Science.* New York, Dover Publications.

Gosset, W. S. (1908). The probable error of a mean. *Biometrica* **6**, 1.

Hald, A. (2007). *A History of Parametric Statistical Inference from Bernoulli to Fisher, 1713–1935.* New York, Springer.

Hammersley, J. M. and Handscomb, D. C. (1964). *Monte Carlo Methods.* London, Chapman and Hall.

Hess, B. (2002). Determining the shear viscosity of model liquids from molecular dynamics simulations. *J. Chem. Phys.* **116**, 209–217.

Huber, P. J. and Ronchetti, E. M. (2009). *Robust Statistics.* 2nd edn., Hoboken, NJ, Wiley.

Huff, D. (1973). *How to Lie with Statistics.* Harmondsworth, Penguin Books.

Jeffreys, H. (1939). *Theory of Probability.* Oxford, Oxford University Press.

Lee, P. M. (1989). *Bayesian Statistics: An Introduction.* New York, Oxford University Press.

Petruccelli, J. Nandram, B. and Chen, M. (1999). *Applied Statistics for Engineers and Scientists.* Upper Saddle River, NJ, Prentice Hall.

Press, W. H., Teukolsky, A. A., Vetterling, W. T. and Flannery, B. P. (1992). *Numerical Recipes, The Art of Scientific Computing.* 2nd edn., Cambridge, Cambridge University Press.

Price, N. C. and Dwek, R. A. (1979). *Principles and Problems in Physical Chemistry for Biochemists.* 2nd edn., Oxford Press, Clarendon Press.

Skyrms, B. (1966). *Choice and Chance.* Belmont, Cal., Wadsworth Publishing.

Straatsma, T. P., Berendsen, H. J. C. and Stam, A. J. (1986). Estimation of statistical errors in molecular simulation calculations. *Mol. Phys.* **57**, 89.

Taylor, J. R. (1997). *An Introduction to Error Analysis. The Study of Uncertainties in Physical Measurements*, 2nd edn. (first edn. 1982). Sausalito, Cal., University Science Books.

Van Kampen, N. G. (1981). *Stochastic Processes in Physics and Chemistry.* Amsterdam, North-Holland.

Walpole, R. E., Myers, R. H., Myers, S. L. and Ye, K. (2007). *Probability and Statistics for Engineers and Scientists.* 8th rev. edn., Upper Saddle River, NJ, Prentice Hall.

Wolter, K. M. (2007). *Introduction to Variance Estimation.* New York, Springer.

Answers to exercises

2.1 (a) $l = 31.3 \pm 0.2$ m (unless the precision is really 20 ± 1 cm; in that case $l = 31,30 \pm 0,20$ m); (b) $c = 15.3 \pm 0.1$ mM; (c) $\kappa = 252$ S/m; (d) $k/\text{L mol}^{-1}\,\text{s}^{-1} = (35.7 \pm 0.7) \times 10^2$ or $k = (35.7 \pm 0.7) \times 10^2$ $\text{L mol}^{-1}\,\text{s}^{-1}$; (e) $= 2.00 \pm 0.03$.

2.2 (a) 173 Pa; (b) 2.31×10^5 Pa $= 2.31$ bar; (c) 2.3 mmol/L; (d) 0.145 nm or 145 pm; (e) 24.0 kJ/mol; (f) 8400 kJ (note that often cal or Cal is written while kcal is meant); (g) 556 N; (h) 2.0×10^{-4} Gy; (i) 0.080 L/km or 8.0 L/100 km; (j) 6.17×10^{-30} Cm; (k) 1.602×10^{-40} F m^2.

3.1 (a) 3.00 ± 0.06 (relative uncertainty 2%); (b) 6.0 ± 0.3 (relative uncertainty $\sqrt{3^2 + 4^2}$%); (c) 3.000 ± 0.001. Note that $\log_{10}(1 \pm \delta) = \pm 0.434 \ln(1 + \delta) \approx \pm 0.434\delta = 0.00087$. Sometimes it is easier to evaluate both boundaries: $\log_{10} 998 = 2.99913$ and $\log_{10} 1002 = 3.00087$; (d) 2.71 ± 0.06 (relative uncertainty $\sqrt{1.5^2 + 1^2}$%).

3.2 $k = \ln 2/\tau_{1/2}$. The relative uncertainty in k equals the relative uncertainty in $\tau_{1/2}$. The absolute uncertainty in $\ln k$ equals the relative uncertainty in $k : \sigma(\ln k) = \sigma(k)/k$. The following values are obtained:

$\frac{1000}{T/\text{K}}$	k/s^{-1}	$\ln(k/\text{s}^{-1})$
1.2771	$(0.347 \pm 0.017) \times 10^{-3}$	-7.97 ± 0.05
1.2300	$(1.155 \pm 0.077) \times 10^{-3}$	-6.76 ± 0.07
1.1862	$(2.89 \pm 0.24) \times 10^{-3}$	-5.85 ± 0.08
1.1455	$(7.70 \pm 0.86) \times 10^{-3}$	-4.87 ± 0.11

Python code for logarithmic plot:
```
autoplotp([Tinv,k],yscale='log',ybars=sigk), with
Tinv, k, sigk from table.
```

3.3 9.80 ± 0.03 (Relative uncertainty is $\sqrt{0.2^2 + (2 \times 0.1^2)} = 0.28\%$)

3.4 Because $\Delta G = RT \ln(kh/k_B T)$, the derivative with respect to T equals $(\Delta G/T) + R$. That is $(30\,000/300) + 8.3 = 108.3$. This implies that a deviation in T of ± 5 yields a deviation in ΔG of $108.3 \times 5 = 540$ J/mol.

3.5 The volume from $r = 1$ equals 4.19 mm^3; the mean of 1000 samples was found to be 4.30 mm^3 and the standard deviation was found to be 1.27. The systematic error in the "naive" volume is -0.11, much less than the standard deviation.

4.1 $f(0) = 0.598\,74$; $f(1) = 0.315\,12$; $f(2) = 0.074\,635$; $f(3) = 0.010\,475$; $f(4) = 0.000\,965$.

4.2 You are looking for $1 - f(0) = 1 - 0.99^{20} = 0.182$.

4.3 With a sample size of n and probability p of voting candidate no. 1, the average number of votes for no. 1 will be pn with variance $p(1 - p)n$ (binomial distribution). To obtain a relative standard deviation of 0.01, $n \geq 10\,000$ is required.

4.4 This distribution is binomial. (a) $\hat{p}_1 = k_0/n$; (b) $\sigma_0 = \sqrt{(k_0 k_1/n)}$; (c) same as (b); (d) Note that deviations in k_0 and k_1 are fully anticorrelated. Therefore $(k_1 \pm \sigma)/(k_0 \mp \sigma) = r(1 \pm \sigma k_1^{-1})/(1 \mp \sigma k_0^{-1}) = r[1 \pm \sigma(k_1^{-1} + k_0^{-1})]$. Standard deviation of r equals $[1 + (k_1/k_0)]/\sqrt{n}$).

4.5 Sum $\mu^k/k!$ over $k = 0$ to $k = \infty$, yielding e^μ.

4.6 Generate Poisson probabilities $f(k, \mu)$ and cumulative probabilities $F(k, \mu)$ from

```
from scipy import stats
f=stats.poisson.pmf
F=stats.poisson.cdf
```

(a) 2.98; (b) $(k \geq 8) : 1 - F(7, 3) = 0.012$; (c) 4 beds; 0.185 patients transported. The optimization can best be done by defining a function `cost(n)`, which computes the costs with n beds, and finding a whole number n for which cost(n) is minimal. For example:

```
def cost(n):
    krange=arange(1,n,1)
    avbeds=(f(krange,3)*krange).sum()+n*
        (1-F((n-1),3))
    return (1-F(n,3))*1500.+(n-avbeds)*300.
```

4.7 This is a Poisson process: s.d. equals the square root of the number of observed impulses. The light measurement gives 900 ± 30 impulses and the dark measurement gives 100 ± 10 impulses. The light intensity is proportional to $(900 - 100) \pm \sqrt{30^2 + 10^2} = 800 \pm 32$. Hence the relative s.d. is 4%. After repeating the measurement 100 times (or after a hundredfold increase of measuring time), the measured numbers become 100× larger, but the (absolute) errors become only 10× larger. The relative uncertainty becomes 10× smaller (0.4%).

4.8 $F(0.1) - F(-0.1) = 2 \times (0.5 - 0.4602) = 0.0796$. Note that this is almost equal to $f(0) \times 0.2 = 0.0798$.

4.9 $f(6) = 6.076 \times 10^{-9}$; $F(-6) = 1.0126 \times 10^{-9}(37/38 + \ldots) = 9.8600 \times 10^{-10}$. Compare to the exact value `stats.norm.cdf(-6.)=` 9.8659×10^{-10}.

4.10 (a) The uniform distribution $f(x) = 1$, $0 \leq x < 1$, has average 0.5 and variance $\sigma^2 = \int_0^1 (x - 0.5)^2 \, dx = 1/12$; adding 12 numbers yields a 12 times larger variance. (b) and (c) with Python code:

```
x=randn(100)
autoplotc(x,yscale='prob')
```

4.11 mean: $\langle t \rangle = 1/k$; variance $\langle ((t - k^{-1})^2 \rangle = 1/k^2$. Use $\int_0^\infty t^n \exp(-kt) \, dt = n!/k^{n+1}$ for evaluating integrals.

4.12 SSR $= 115.6$; SSE $= 154.0$; F $= 6.005$; cdf(F, 1, 8) $= 0.96$; treatment is significant at 5% confidence level.

5.1 Yes, Fig. 2.1 gives a straight line; $\mu = 8.68$; $\sigma = 1.10$. Accuracy ca 0.05.

5.2 Just work out the square in $\frac{1}{n} \sum (x_i - \langle x \rangle)^2$.

5.3 No: apply the equation to $y = x - c$; all terms with c cancel.

5.4 Usually things go wrong for c exceeding 10^7. Suggestion: use Python function:

```
def rmsd(c):
    n=1000
    x=randn(n)+c
    xav=x.sum()/n
    rmsd1=((x-xav)**2).sum()/n
    rmsd2=(x**2).sum()/n - xav**2
    return [rmsd1,rmsd2]
```

The first value is correct; the second may be in error.

5.5 The estimated s.d. equals $\hat{\sigma} = \sqrt{\langle (\Delta x)^2 \rangle n/(n-1)}$, where $\langle (\Delta x)^2 \rangle$ is the mean squared deviation. For $n = 15$ the s.d. in σ is 19%; this gives $\hat{\sigma} = 5 \pm 1$. For $n = 200$ the s.d. in σ is 5%; this gives $\hat{\sigma} = 5.1 \pm 0.3$. In the first case the mean is 75 ± 5; in the second case the mean is 75.3 ± 5.1.

5.6 1. (a) average: 29.172 s; (b) msd: $0.0315\,s^2$; (c) rmsd: 0.1775 s; (d) range: 28.89–29.43 s; median: 29.24 s; first quartile: 29.02 s; third quartile: 29.33 s

2. (a) mean: 29.172 s; (b) variance: $0.0354\,s^2$; (c) s.d.: 0.188 s; (d): 0.063 s; (e) 0.0177 s; 0.047 s; 0.016 s.

3. 29.16 ± 0.06 km/hr; deviation: $+6.6 \pm 4\%$ km/hr.

4. No. The inaccuracy of keeping the right speed is incorporated into the measurements.

5. 80%: 29.10–29.25; 90%: 29.07–29.27; 95%: 29.06–29.28 s.

6. 80%: 123.06–123.74; 90%: 123.00–123.82; 95%: 122.91–123.92 km/hr.

7. 80%: 123.06–123.76; 90%: 122.97–123.85; 95%: 122.88–123.94 km/hr.

8. 80%: 123.03–123.76; 90%: 122.91–123.90; 95%: 122.79–124.02 km/hr.

5.7 Use weighted averaging: $N_A = 6.022\,141\,89(20)$.

5.8 The plot can be made by first constructing a list z of all 27 possible values:

```
z=[-1.]+[-2./3.]*3+[-1./3.]*6+[0.]*7+[1./3.]*6
  +[2./3.]*3+[1.]
autoplotc(z,yscale='prob')
```

This plot perfectly fits a straight line through (0, 50%); $\sigma = 0.47$ (exact: 0.471).

5.9 Note that the characteristic function of $\delta(x - a)$ equals $\exp(iat)$. The probability density function of a variable x, randomly chosen from -1, 0 and $+1$, consists of three delta functions $\Phi(t) = \frac{1}{3}\delta(x+1) + \frac{1}{3}\delta(x) + \frac{1}{3}\delta(x - 1)$. Its characteristic function is $\frac{1}{3}[1 + \exp(-it) + \exp(it)]$. The pdf of the sum of three such variables x_1, x_2, x_3 is the convolution of

$f(x_1), f(x_2)$ and $f(x_3)$; its characteristic function equals $\Phi(t)^3$. Working out the third power yields

$[\exp(3it) + 3\exp(2it) + 6\exp(-it) + 7 + 6\exp(-it) + 3\exp(-2it) + \exp(-3it)]/27$.

Its Fourier transform contains seven delta functions at $x = -3, -2, -1, 0, 1, 2, 3$. If not the sum but the average of three values is taken, the x values reduce by a factor 3.

The variance can be obtained from the second derivative of the characteristic function at $t = 0$, or directly from the pdf, and equals 2 for the sum, or 2/9 for the average.

6.1 Line goes through points (9, 100) and (188, 1) (precision ca 1%). Gives $k = \ln 100/(188 - 9) = 0.0257$ and $c_0 = 126$.

6.2 (In too many decimals:) Lineweaver–Burk: $K_m = 1/0.0094 = 106.383$; $v_{max} = K_m(0.04 + 0.0094)/0.35 = 15.015$; Eadie–Hofstee: $K_m = (15 - 2)/(0.120 - 0.007) = 115.04$; $v_{max} = 0.120 K_m + 2 = 15.805$; Hanes: $v_{max} = 500/(39 - 7.5) = 15.873$; $K_m = 7.5 v_{max} = 119.05$.

6.3 Plot the data $1000/T, k$ on a horizontal scale from 1.14 to 1.30. Draw the best line through the points; this line goes through $(1.14, 9.5e - 3)$ and $(1.30, 2.0e - 4)$. Hence $E/1000R = [\ln(9.5e - 3/2.e - 4)]/[1.30 - 1.14] = 24.13$ and $E = 200.63$ kJ/mol. Varying the slope yields E between 191.69 and 208.24. Result: $E = 201 \pm 8$ kJ/mol. Your values may differ (insignificantly) from these numbers.

6.4 68.8 ± 0.6 mmol/L (note that the unit molar (M, mol/L) is obsolete).

7.1 Use the Python program fit (code **7.7**). With the function $y = ax + b$, the best fit gives $a = 7.23 \pm 0.31$ and $b = 0.0636 \pm 0.0017$, with correlation coefficient $\rho_{ab} = -0.816$. From this follows $v_{max} = 1/b = 15.7 \pm 0.4$ and $K_m = a/b = 114 \pm 8$. The *relative* uncertainty δ in a/b is found from

$$\delta^2 = \left(\frac{\sigma_a}{a}\right)^2 + \left(\frac{\sigma_b}{b}\right)^2 - 2\rho_{ab}\frac{\sigma_a \sigma_b}{ab}.$$

A direct nonlinear fit to the data [S, v] yields $v_{max} = 15.7 \pm 0.4$ and $K_m = a/b = 115 \pm 8$.

7.2 Use the Python program fit (code **7.7**). With the function $y = -aT + b$ you find $\Delta S = a = 0.259 \pm 0.013$, $b = 110.3 \pm 3.9$ and $\rho_{ab} = 0.99778516$. Extrapolation to $T = 350$ gives $\Delta G(350) = 19.81 \pm 0.71$, where the s.d. has been calculated from

$$\sigma_{\Delta G}^2 = 350^2 \sigma_a^2 + \sigma_b^2 - 2.350.\rho_{ab}\sigma_a\sigma_b.$$

With the function $y = -a(T - 300) + b$ you find $\Delta S = a = 0.259 \pm 0.013$, $b = 32.74 \pm 0.26$ and $\rho_{ab} = 0$. Extrapolation to $T = 350$ now gives $\Delta G(350) = 19.81 \pm 0.71$, where the s.d. has been calculated from

$$\sigma_{\Delta G}^2 = 50^2 \sigma_a^2 + \sigma_b^2.$$

The results are exactly the same, but the extrapolation is much simpler in the second case where $\rho = 0$.

7.3

$$\sigma_y^2 = \left(\frac{dy}{dt}\right)^2 \sigma_t^2 = \frac{\sigma_t^2}{t^2}.$$

Hence $w_i = \sigma_y^{-2} = t_i^2/\sigma_t^2 \propto t_i^2$.

7.4 $a = 71.5 \pm 3.8$; $b = 19.1 \pm 3.9$; $p = 0.0981 \pm 0.0061$; $q = 0.0183 \pm 0.0034$. Note that these values deviate from the graphical estimate. Fitting to multiple exponentials is quite difficult; the parameters have a strong mutual correlation (e.g. $\rho_{ab} = 0.98$) and sometimes a minimum cannot be found.

7.5 For $c =$ position lens, $yf(x, [f, c]) = c + f * (c - x)/(c - x - f)$. Least-squares fitting yields $f = 55.15$; $c = 187.20$. The $S_0 = 3.13$; 4 degrees of freedom. Covariance matrix ($S_0/4*$ leastsq output yields $\sigma_1 = 0.2$; $\sigma_2 = 0.3$; $\rho = 0.91$. Result: $f = 55.1 \pm 0.2$ mm.

7.6 Find out by yourself.

7.7 The output of the program report gives sufficient comment. Try
`x=arange(100.); sig=ones(100)`
`y1=randn(100); y2=y1+0.01*x`
`report([x,y1,sig])` may produce an insignificant drift, while y2 may imply a significant drift.

Appendices

Why do squared uncertainties add up in sums?

We wish to determine the sum $f = x + y$ of two quantities, each of which are drawn from a probability distribution, with

$$E[x] = \mu_x; \quad E[(x - \mu_x)^2] = \sigma_x^2, \tag{A1.1}$$

$$E[y] = \mu_y; \quad E[(y - \mu_y)^2] = \sigma_y^2. \tag{A1.2}$$

The quantity $f = x + y$ has the expectation

$$\mu = \mu_x + \mu_y \tag{A1.3}$$

and a variance

$$\sigma_f^2 = E[(f - \mu)^2] = E[(x - \mu_x + y - \mu_y)^2]$$
$$= E[(x - \mu_x)^2 + (y - \mu_y)^2 + 2(x - \mu_x)(y - \mu_y)]$$
$$= \sigma_x^2 + \sigma_y^2 + 2E[(x - \mu_x)(y - \mu_y)]. \tag{A1.4}$$

If x and y are independent of each other (i.e., the deviations from the means of x and y are statistically independent samples), then the last term vanishes.[1] In that case the squared uncertainties (the variances) indeed add up to yield the squared uncertainty of the sum.

From the derivation we see immediately that squared uncertainties no longer simply add up when the deviations of the two contributing quantities are correlated. The quantity $E[(x - \mu_x)(y - \mu_y)]$ is the *covariance* of x and y. The covariance is often expressed relative to the variances themselves as the *correlation coefficient* ρ_{xy}:

$$\text{cov}(x, y) = E[(x - \mu_x)(y - \mu_y)] \tag{A1.5}$$

$$\rho_{xy} = \frac{\text{cov}(x, y)}{\sigma_x \sigma_y}. \tag{A1.6}$$

---

[1] Strictly, the requirement is that the two quantities are uncorrelated, i.e., their covariance is zero. This is a less severe requirement than being independent.

The complete equation for a sum thus is

$$\operatorname{var}(x+y) = \operatorname{var}(x) + \operatorname{var}(y) + 2\operatorname{cov}(x,y). \qquad (A1.7)$$

For a *difference* $f = x - y$ the equation is

$$\operatorname{var}(x-y) = \operatorname{var}(x) + \operatorname{var}(y) - 2\operatorname{cov}(x,y). \qquad (A1.8)$$

For a product, resp. a quotient, the same equations are valid for the *relative* deviations:

$$\frac{\operatorname{var}f}{f^2} = \frac{\operatorname{var}x}{x^2} + \frac{\operatorname{var}y}{y^2} \pm 2\frac{\operatorname{cov}(x,y)}{xy}, \qquad (A1.9)$$

where the plus sign is valid for $f = xy$ and the minus sign for $f = x/y$.
    The general equation for the variance of a function $f(x_1, x_2, \ldots)$ is

$$\operatorname{var}(f) = \sum_{i,j} \frac{\partial f}{\partial x_i} \frac{\partial f}{\partial x_j} \operatorname{cov}(x_i, x_j). \qquad (A1.10)$$

Here $\operatorname{cov}(x_i, x_i) = \operatorname{var}(x_i)$. This equation follows directly by taking the square of

$$df = \sum_i \frac{\partial f}{\partial x_i} dx_i.$$

The assumption is made that deviations are small, so that only the first order in a Taylor expansion need be considered.
    Here is an example of the use of covariances. Suppose we have performed (with a computer program) a least-squares analysis of $f(x) = ax + b$ on a number of data points with the result:[2]

$$a = 2.30526; \quad b = 5.21632;$$

$$\sigma_a = 0.00312; \quad \sigma_b = 0.0357; \quad \rho_{ab} = 0.7326.$$

These results will be used for an inter- or extrapolation: *What value and standard deviation is expected for $f(10)$?*
    For this purpose we first determine the values, variances and covariances for the two quantities ax and b we want to add. In this case x is a multiplying

---

[2] Note that the numbers are given in too many digits. This is good practice for intermediate results of a statistical analysis, since unnecessary rounding errors are thus avoided.

factor that appears quadratically in var (ax) and linearly in cov (ax, b):

$$ax = 23.0526; \quad b = 5.21632; \quad f = 28.26892;$$

$$\text{var}\,(ax) = 0.00312^2 \times 10^2; \quad \text{var}\,(b) = 0.0357^2;$$

$$\text{cov}\,(ax, b) = 10 \times 0.7326 \times 0.00312 \times 0.0357.$$

Insertion in Eq. (A1.7) then gives var $(f) = 0.00388$. Had we disregarded the covariance, var (f) would have appeared to be equal to 0.00225. The s.d. of f now is 0.0623 and we give the result as $f = 28.27 \pm 0.06$.

A2 Systematic deviations due to random errors

When $f(x)$ has a non-negligible curvature, systematic deviations may occur in f as a result of random deviations in x, even when the latter are symmetrically distributed. Suppose you have a batch of spheres with approximately – but not exactly – equal radii. You measure the radii and find an approximately normal distribution with $r = 1.0 \pm 0.1$ mm. For the volume you thus find (in too many decimals): $V = \frac{4}{3}\pi r^3 = 4.19$ mm^3. However, if you work out the third power of r to higher order, you find:

$$(r \pm \Delta r)^3 = r^3 \pm 3r^2 \Delta r + 3r(\Delta r)^2 \pm (\Delta r)^3.$$

Assuming the distribution function for Δr to be symmetric, you must conclude that the third term is always positive and thus gives a contribution to the expected value of f:

$$E[r^3] = r^3 + 3r\,\text{var}\,(r).$$

If $E[f(x)] \neq f(E[x])$ we have a *systematic deviation* or *bias*. In our example this extra contribution to the volume is 0.13 mm^3 and the expected volume is 4.32 mm^3. Without this correction, the predicted volume has a bias of -0.13. This is ten times smaller than the standard deviation itself and is therefore not very important. But there are cases when this kind of bias must be corrected.

The general equation results from the second term in a Taylor expansion:

$$f(x) = f(a) + (x - a)f'(a) + \frac{1}{2}(x - a)^2 f''(a) + \cdots \qquad \text{(A2.1)}$$

$$E[f] = f(E[x]) + \frac{1}{2}\frac{d^2 f}{dx^2}\,\text{var}\,(x) + \cdots \qquad \text{(A2.2)}$$

A special case: sampling exponential functions

There is at least one type of fairly common application where evaluation of the bias is essential: computing the average over an exponential function

138

of a statistically distributed observable. For example, computation of the thermodynamic potential μ of a molecular species in a molecular simulation (molecular dynamics or Monte Carlo) by the *particle insertion method* requires many random trial insertions of a particle. If the computed interaction energy of the inserted particle with its environment of the i-th insertion is E_i, the *excess* thermodynamic potential (in excess of the ideal gas value) is approximated by

$$\mu^{\text{exc}} = \beta^{-1} \ln \left[\frac{1}{N} \sum_{i=1}^{N} e^{-\beta E_i} \right], \tag{A2.3}$$

where $\beta = 1/(k_B T)$ with k_B = Boltzmann's constant and T the absolute temperature. The same kind of averaging occurs in other types of free-energy determinations from simulations. The reader is referred to Berendsen (2007)[1] for details on the physics of these methods.

The essential statistics in problems of this kind can be formulated as *averaging over an exponential function* of a randomly sampled variable x, with distribution function $f(x)$. We are interested in the logarithm of such an average:

$$y = -\frac{1}{\beta} \ln \langle e^{-\beta x} \rangle, \tag{A2.4}$$

where

$$\langle e^{-\beta x} \rangle = E[e^{-\beta x}] = \int_{-\infty}^{\infty} f(x) e^{-\beta x} \, dx. \tag{A2.5}$$

The parameter β functions as a scaling for x: given a fixed probability distribution for x, the larger β, the more severe the statistical problems on averaging appear to be. The problem is that occasional large negative values for x contribute heavily to the average. We can get some insight by expanding y in powers of β; such an expansion is called a *cumulant expansion*. For simplicity we take $\langle x \rangle = 0$, so that all moments of the distribution of x are central moments. Adding an arbitrary value a to every x simply results in adding a to the result y. The cumulant expansion is

$$y = -\frac{\beta}{2!} \langle x^2 \rangle + \frac{\beta^2}{3!} \langle x^3 \rangle - \frac{\beta^3}{4!} (\langle x^4 \rangle - 3 \langle x^2 \rangle^2) + \mathcal{O}(\beta^4). \tag{A2.6}$$

For a normal distribution only the first term survives, resulting in

$$y = -\beta/2, \tag{A2.7}$$

as can be checked by direct integration of (A2.5). This is a bias due to the width of the normal distribution. If we know for sure that the distribution

[1] See reference list on page 123.

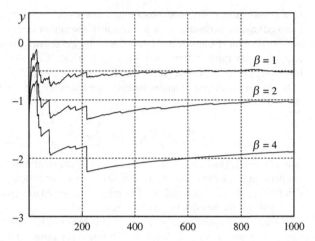

Figure A2.1 Cumulative average of $y = -\beta^{-1} \ln\langle\exp(-\beta x)\rangle$ over n samples drawn from a normal distribution (average 0, variance $\sigma^2 = 1$). The theoretical limits are -0.5β, indicated by dotted lines (from Berendsen, 2007).

function of x is normal, y can be accurately determined from (A2.7), but it is difficult to determine y from random sampling of x. To show this, Fig. A2.1 gives the values of y obtained from running averages for 1000 samples of x from a normal distribution and for three values of β. It turns out that 1000 samples are barely sufficient to find convergence for $\beta = 2$, but for $\beta = 4$ this number by no means suffices.

A3 Characteristic function

The characteristic function $\Phi(t)$ of a probability density function $f(x)$:

$$\Phi(t) \stackrel{\text{def}}{=} E[e^{itx}] = \int_{-\infty}^{\infty} e^{itx} f(x)\, dx \qquad \text{(A3.1)}$$

has some interesting properties. In fact, $\Phi(t)$ is the Fourier transform of $f(x)$. This implies that the characteristic function of the *convolution* $f_1 * f_2$ of two density functions f_1 and f_2 is the *product* of the two corresponding characteristic functions Φ_1 and Φ_2. The convolution, defined by

$$f_1 * f_2(x) = \int_{-\infty}^{\infty} f_1(x - \xi) f_2(\xi)\, d\xi, \qquad \text{(A3.2)}$$

is the density distribution of the *sum* of two random variables $x_1 + x_2$ when the density functions of x_1 and x_2 are resp. f_1 and f_2. The *convolution theorem* of Fourier analysis states that the Fourier transform of a convolution equals the product of the Fourier transforms of the contributing terms. This product rule also applies to convolutions of n functions.

Another interesting property of the characteristic function is that its series expansion in powers of t generates the *moments* of the distribution. The characteristic function is therefore often called the *moment-generating function*. Since

$$e^{itx} = \sum_{n=0}^{\infty} \frac{(itx)^n}{n!}, \qquad \text{(A3.3)}$$

it follows that

$$\Phi(x) = E[e^{itx}] = \sum_{n=0}^{\infty} \frac{(it)^n}{n!} E[x^n] = \sum_{n=0}^{\infty} \frac{(it)^n}{n!} \mu_n. \qquad \text{(A3.4)}$$

The moments are also given by the *derivatives* of the characteristic function at $t = 0$:

$$\Phi^{(n)}(0) = \frac{d^n \Phi}{dt^n}\Big|_{t=0} = i^n \mu_n. \qquad \text{(A3.5)}$$

141

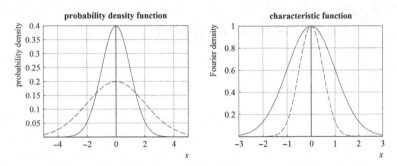

Figure A3.1 Left: a probability density function (in this case a normal distribution); right: its characteristic function. The dashed curve has a standard deviation twice that of the drawn curve.

The μ_n are the moments, not the central moments. But you can always choose the origin of x at the position of the mean.

A special case is the variance σ^2:

$$\sigma^2 = -\frac{d^2\Phi}{dt^2}(0). \qquad (A3.6)$$

Figure A3.1 shows the relation between a density function and its characteristic function. The density function is normalized by its integral; the characteristic function is always equal to 1 for $t = 0$. The broader the density function, the narrower the characteristic function.

A4 From binomial to normal distributions

A4.1 The binomial distribution

Consider a case where the outcome of an observation x can be either 0 or 1 (or tail or head, or no or yes, or false or true, or absent or present, or whatever binary choice you wish to define). Let the probability of obtaining 1 be equal to p, meaning $E[x] = p$. Then for two observations the following combinations may occur: 00, 01, 10, 11. Assuming that successive observations are independent, the probability $f(k)(k = 0, 1, 2)$ that exactly k times a 1 is observed is

$$f(0) = (1 - p)^2$$
$$f(1) = 2p(1 - p)$$
$$f(2) = p^2. \tag{A4.1}$$

In general: the probability $f(k; n)$ that in n independent observations exactly k times a 1 is observed equals

$$f(k; n) = \binom{n}{k} p^k (1 - p)^{(n-k)}, \tag{A4.2}$$

where

$$\binom{n}{k} = \frac{n!}{k!(n - k)!} \tag{A4.3}$$

is the *binomial n over k*, i.e., the number of ways k items can be chosen from a collection of n items. For the case considered above: $n = 2$, the three binomial coefficients ($k = 0, 1, 2$) are respectively 1, 2 and 1; these are the coefficients in $f(k)$ of (A4.1). This is the *binomial distribution*.

Note that the sum of all probabilities equals 1:

$$\sum_{k=0}^{n} f(k; n) = \sum_{k=0}^{n} \binom{n}{k} p^k (1 - p)^{(n-k)} = (p + 1 - p)^n = 1. \tag{A4.4}$$

143

The mean $E[k]$ is defined by the sum

$$E[k] = \sum_{k=0}^{n} kf(k;n);$$ (A4.5)

this can be worked out as

$$E[k] = pn \sum_{k=1}^{n} \binom{n-1}{k-1} p^{k-1}(1-p)^{\{n-1-(k-1)\}} = pn.$$ (A4.6)

Similarly (details are left to the reader), the variance equals

$$E[(k-pn)^2] = E[k^2] - 2pnE[k] + (pn)^2$$

$$= \sum_{k=1}^{n} k^2 f(k;n) - (pn)^2 = p(1-p)n.$$ (A4.7)

A4.2 The multinomial distribution

When there are more than one (e.g. m) possible values for the sampled variable, with probabilities $p_1, p_2, \ldots, p_m$ ($\sum_i p_i = 1$), then the distribution is a *multinomial distribution*:

$$f(k_1, k_2, \ldots, k_m; n) = \frac{n!}{k_1! k_2! \ldots k_m!} \Pi_{i=1}^{m} p_i^{k_i}; \quad \sum_i k_i = n.$$ (A4.8)

This is an example of a multidimensional *joint* probability, meaning the probability that event 1 occurs k_1 times *and* event 2 occurs k_2 times *and* etc. The means and variances for each of the number of occurrences are the same as for the binomial distribution:

$$E[k_i] = \mu_i = np_i,$$ (A4.9)

$$E[(k_i - \mu_i)^2] = \sigma_i^2 = np_i(1 - p_i).$$ (A4.10)

The fact that the sum of all k_i is constrained causes a covariance between k_i and $k_j (i \neq j)$:

$$covar(k_i, k_j) = E[(k_i - \mu_i)(k_j - \mu_j)] = -np_i p_j.$$ (A4.11)

The covariance matrix is a symmetric matrix of which the diagonal elements are the variances and the non-diagonal elements are the covariances.

A4.3 The Poisson distribution

From binomial to Poisson

Consider a suspension of small particles. You want to determine the average number of particles per unit volume by counting the number of particles under a microscope in a sample of $0.1 \times 0.1 \times 0.1$ mm (10^{-6} cm^3). If the number density is known, and thus the average number of particles in the sample volume is known, what then is the probability of finding exactly k particles in the small volume?

Let the average number of particles in the sample volume be μ. Divide the sample volume into a large number n of cells, small enough to contain no more than one particle. The probability that a specified cell contains a particle equals $p = \mu/n$. The probability that precisely k particles will be found in the sample volume equals the binomial distribution $f(k; n)$ with $p = \mu/n$.

Equivalently you may consider another example: Electrical impulses (or photons, or gamma quanta, or any other short events) occur randomly and independently of each other. You observe the events during a given time span T. If the average number of events within a time T is known, what then is the probability that precisely k events are observed in a time span of length T? In this case we divide the interval T into n short time intervals. Let the average number of events in a time T be μ. The probability that precisely k events will be counted in the interval T equals the binomial distribution $f(k; n)$ with $p = \mu/n$.

Now let the number of cells, or the number of time intervals, n go to infinity, while $pn = \mu$ is kept constant. This means that $p \to 0$, but in such a way that $pn = \mu$ remains the same. Thus $k \ll n$. The binomial coefficient then approaches $n^k/k!$:

$$\frac{n!}{k!(n-k)!} = \frac{n(n-1)\ldots(n-k+1)}{k!} \approx \frac{n^k}{k!}, \tag{A4.12}$$

so that

$$p(k) \to \frac{n^k}{k!} \left(\frac{\mu}{n}\right)^k \left(1 - \frac{\mu}{n}\right)^{n-k}.$$

The term on the right approaches $e^{-\mu}$ because $n - k \to n$ and

$$\lim_{n\to\infty} \left(1 - \frac{\mu}{n}\right)^n = e^{-\mu}, \tag{A4.13}$$

from which it follows that

$$f(k) = \frac{\mu^k e^{-\mu}}{k!}. \tag{A4.14}$$

This is the probability mass function of the *Poisson distribution* for k, given the average μ.

The Poisson distribution is a discrete distribution: the observed number k can only assume positive integer values $0, 1, 2, \ldots$ The mean μ is a parameter of the distribution and can be any positive real number.

Properties of the Poisson distribution

It is easy to show that the Poisson distribution is normalized and that its mean equals μ. Prove this by using the series expansion

$$e^\mu = \sum_{k=0}^{\infty} \frac{\mu^k}{k!}. \tag{A4.15}$$

The variance of the distribution is

$$\text{var}(k) = \sigma^2 = E[(k - \mu)^2] = \mu. \tag{A4.16}$$

This follows from $\sum_{k=0}^{\infty} k^2 \mu^k / k! = \mu^2 + \mu$, but it is also the limit of (A4.10) for $p \to 0$.

A4.4 The normal distribution

From Poisson to normal

For large values of μ the Poisson distribution approaches a normal distribution with mean μ and s.d. $\sqrt{\mu}$. When we attempt to derive this limit we must be very careful to retain a sufficiently high order in the approximations as terms tend to compensate each other.

Let both k and μ go simultaneously to ∞, but in a coordinated way. Define

$$x = \frac{k - \mu}{\sqrt{\mu}}; \quad k = \mu + x\sqrt{\mu},$$

and use the Stirling approximation of the factorial $k!$:

$$k! = k^k e^{-k} \sqrt{2\pi k} [1 + O(k^{-1})]. \tag{A4.17}$$

The logarithm of the Poisson probability (A4.14) expands in orders of k^{-1} as follows:

$$\ln f(k) = k - \mu - k \ln(k/\mu) - \frac{1}{2} \ln(2\pi k) + O(k^{-1})$$

$$= x\sqrt{\mu} - (\mu + x\sqrt{\mu}) \ln\left(1 + \frac{x}{\sqrt{\mu}}\right) - \frac{1}{2} \ln\left[2\pi\mu\left(1 + \frac{x}{\sqrt{\mu}}\right)\right].$$

Because $\ln \mu \to \infty$, the whole expression goes to $-\infty$! This is what we expect because we calculate the probability of finding precisely one (integer) value of k (which obviously goes to zero) and *not* the probability density of $f(x)$. When we expand the logarithm

$$\ln(1 + z) = z - \frac{1}{2}z^2 + O(z^3),\qquad\qquad (A4.18)$$

we find eventually that

$$\lim_{k \to \infty} \ln f(k) = -\frac{1}{2}x^2 - \frac{1}{2}\ln(2\pi\mu).$$

The distance between two successive discrete values of x is

$$\Delta x = \frac{k + 1 - \mu}{\sqrt{\mu}} - \frac{k - \mu}{\sqrt{\mu}} = \frac{1}{\sqrt{\mu}};$$

therefore there are $\sqrt{\mu}\, dx$ discrete values between x and $x + dx$. It follows that

$$f(x)\, dx = \frac{1}{\sqrt{2\pi}} \exp\left(-\frac{x^2}{2}\right) dx,\qquad\qquad (A4.19)$$

what we set out to prove.

A5 *Central limit theorem*

While the central limit theorem can be found in almost all textbooks on statistics, its proof is almost never given. The best reference in the accessible literature that includes a discussion of the validity limitations can be found in Cramér (1946).[1] Van Kampen (1981)[2] gives a more intuitive discussion. You will need to know what the characteristic function of a probability distribution is (see Appendix A3 on page 141). Cramér states the central limit theorem as follows:

> *Whatever be the distributions of the independent variables x_i – subject to certain very general conditions – the sum $x = x_i + \cdots + x_n$ is asymptotically normal (m, σ), where m is the sum of means and σ^2 is the sum of variances.*

"Asymptotically normal" means that the distribution of x tends to the normal distribution $N(m, \sigma)$ for large n. The "certain very general conditions" include the requirement that every contributing distribution has a finite variance; in addition the sum of third moments, divided by the 3/2 power of the total variance, must tend to zero for large n. The latter is of course always true for symmetric distributions, but it is also true for a sum of equivalent distributions. It is false only in pathological cases.

Consider a large number n of independent continuous random variables $x_1, x_2, \ldots, x_n$ with sum x:

$$x = \sum_{i=1}^{n} x_i, \tag{A5.1}$$

each with a probability density function $f_i(x)$. Let each pdf have a finite mean m_i and variance σ_i^2. We now ask the question what can we say about the probability density function $f(x)$, when n tends to infinity.

First eliminate the mean. Since

$$\sum_i (x_i - m_i) = \sum_i x_i - \sum_i m_i = x - m, \tag{A5.2}$$

---

[1] See reference list on page 123.
[2] See reference list on page 124.

148

the mean of x is the sum of the means m_i. So by considering $x_i - m_i$ instead of x_i, all contributing variables and the resulting sum have zero mean. Now consider the density function of the sum $f(x)$. This is a *convolution* of all f_i and hence the characteristic function $\Phi(t)$ of $f(x)$ is the *product* of the characteristic functions $\Phi_i(t)$ of $f_i(x_i)$:

$$\Phi(t) = \Pi_{i=1}^{n} \Phi_i(t), \tag{A5.3}$$

or

$$\ln \Phi(t) = \sum_{i=1}^{n} \ln \Phi_i(t), \tag{A5.4}$$

where

$$\Phi_i(t) \stackrel{\text{def}}{=} \int_{-\infty}^{\infty} e^{ixt} f_i(x)\, dx. \tag{A5.5}$$

We know that $\Phi(0) = 1$, but for $t \neq 0$ each $\Phi_i(t) < 1$ (because at $t = 0$ the first derivative is zero and the second derivative is negative) and hence the product $\Phi(t)$ tends to zero. So $\Phi_i(t)$ is a rapidly decaying function of t. How does it behave for small t?

Consider the expansion of $\ln \Phi_i(t)$ in powers of t, which follows from the expansion (A3.4) on page 141 of $\Phi_i(t)$:

$$\ln \Phi_i(t) = -\frac{1}{2}\sigma_i^2 t^2 - \frac{i}{6}\mu_{3i} t^3 + \frac{1}{24}(\mu_{4i} - 3\sigma_i^4)t^4 + \cdots \tag{A5.6}$$

From this we find, denoting $\sum_i \sigma_i^2$ by σ^2:

$$\ln \Phi(t) = -\frac{1}{2}\sigma^2 t^2 \left[1 + \frac{i}{3}\frac{\sum_i \mu_{3i}}{\sigma^3}\sigma t - \frac{1}{12}\left(\frac{\sum_i \mu_{4i}}{\sigma^4} - 3\frac{\sum_i \sigma_i^4}{\sigma^4} \right)\sigma^2 t^2 \cdots \right] \tag{A5.7}$$

Since – under mild conditions – terms such as $\sum_i \sigma_i^2, \sum_i \mu_{3i}$, etc. scale proportional to n, the factor $\sum_i \mu_{3i}/\sigma^3$ in the second term scales as $n^{-1/2}$ and the factor in the third term scales as n^{-1}. So, for large n, $\ln \Phi(t)$ approaches $-\sigma^2 t^2/2$:

$$\lim_{n \to \infty} \Phi(t) = e^{-(\sigma^2 t^2/2)}, \tag{A5.8}$$

which implies that the probability density function is normal:

$$\lim_{n \to \infty} f(x) = \frac{1}{\sigma\sqrt{2\pi}} \exp\left(-\frac{x^2}{2\sigma^2} \right). \tag{A5.9}$$

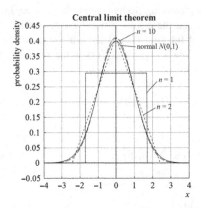
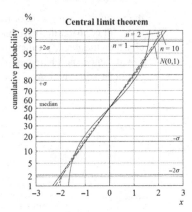

Figure A5.1 The probability distribution of the sum of n random numbers ($n = 1, 2, 10$), chosen from a uniform distribution between $-a$ and $+a$, compared with the normal distribution $N(0, 1)$. Each distribution has unit variance. Left: probability density, right: cumulative distribution on a probability scale.

Summarizing we can say that, with increasing n, the higher powers of t in $\ln \Phi(t)$ die out relative to the t^2-term; the higher the power, the faster it dies out. The most persistent is the third power (related to the skewness), which diminishes only slowly with the inverse square root of n.

An example, related to Exercise 4.10 on page 51, is the distribution function of the sum of n random numbers, sampled from a uniform distribution in the domain $[-a, a\rangle$. Figure A5.1 gives the distribution functions for $n = 1, 2, 10$, compared with the normal distribution $N(0, 1)$. In each case a has been chosen such that the distribution function of the resulting sum variable has a standard deviation of 1.

 See **Python code** A5.1 on page 194 for a Python code to generate the distribution function for arbitrary n using Fourier transforms.

It is clear that the derivation fails completely when one or more of the contributing distributions has an undefined (infinite) variance, such as the Lorentz distribution (see page 43) has. In fact, the sum of Lorentz-distributed random variables remains Lorentz-distributed!

A6 Estimation of the variance

Why is the best estimate for the variance larger than the mean squared deviation of the average?

Assume x_i are independent samples from a distribution $f(\mu, \sigma)$ with mean μ and s.d. σ. In order to find out the relation between $\langle (\Delta x)^2 \rangle$ and σ it is necessary to compute the expectation of $\langle (\Delta x)^2 \rangle$. After realizing that

$$
\begin{aligned}
\langle (\Delta x)^2 \rangle &= \langle (x - \langle x \rangle)^2 \rangle \\
&= \langle [x - \mu - (\langle x \rangle - \mu)]^2 \rangle, \\
&= \langle (x - \mu)^2 \rangle - (\langle x \rangle - \mu)^2,
\end{aligned}
\tag{A6.1}
$$

we see that

$$
E[\langle (\Delta x)^2 \rangle] = \sigma^2 - E\left[\left(\frac{1}{n} \sum_{i=1}^{n} (x_i - \mu) \right)^2 \right]
$$

$$
= \sigma^2 - \frac{1}{n^2} \sum_{i=1}^{n} \sum_{j=1}^{n} E[(x_i - \mu)(x_j - \mu)]
\tag{A6.2}
$$

Uncorrelated data points

When all samples are independent of each other (and therefore uncorrelated),[1] the double sum reduces to a single sum because x_i and x_j are

---

[1] The terms *independent* and *uncorrelated* mean different things. Two random variables x and y are statistically independent when the random processes selecting either of them are independent of each other; x and y are statistically uncorrelated when $E[(x - \mu_x)(y - \mu_y)] = 0$. Independent samples are also uncorrelated, but uncorrelated samples need not be independent. For example, the random variable x sampled from $N(0, 1)$ and x^2 are uncorrelated (because $E[x^3] = 0$), but they are not independent!

151

independent and only the term $j = i$ in the second sum survives:

$$E[\langle(\Delta x)^2\rangle] = \sigma^2 - \frac{1}{n^2}\sum_{i=1}^{n}E[(x_i - \mu)^2]$$

$$= \sigma^2\left(1 - \frac{1}{n}\right). \tag{A6.3}$$

Thus it follows that the best estimate for σ^2 equals $n/(n-1)$ times the average of the squared deviations from the average. Note that this is true for any kind of distribution with finite variance.

Correlated data points

In the derivation of (A6.3) explicit use has been made of the assumption that the deviations from the mean are uncorrelated. In practice subsequent data points are often correlated, i.e., $E[(x_i - \mu)(x_j - \mu)] \neq 0$ for $j \neq i$. If the latter is the case, more terms will remain in the double sum of (A6.2) and a larger term will be subtracted from σ^2. The best estimate for the variance will be larger.

Here we shall give the equation for the cases that a known correlation between *successive* data points exists (Straatsma *et al.*, 1986).[2] The assumption is made that the ordered series $x_1, \ldots, x_n$ is a *stationary stochastic variable*, i.e. it has a constant variance and correlation coefficients between x_i and x_j that only depend on the distance $|j - i|$.

The term with the double sum in (A6.2) is:

$$\frac{1}{n^2}\sum_i\sum_j E[(x_i - \mu)(x_j - \mu)] = \sigma^2\frac{n_c}{n}, \tag{A6.4}$$

where n_c is a kind of *correlation length*:

$$n_c = 1 + 2\sum_{k=1}^{n-1}\left(1 - \frac{k}{n}\right)\rho_k. \tag{A6.5}$$

Here ρ_k is the correlation coefficient between x_i and x_{i+k}:

$$E[(x_i - \mu)(x_{i+k} - \mu)] = \rho_k\sigma^2. \tag{A6.6}$$

[2] See reference list on page 124.

$i \rightarrow$

$j \downarrow$

1	ρ_1	ρ_2	ρ_3	ρ_4
ρ_1	1	ρ_1	ρ_2	ρ_3
ρ_2	ρ_1	1	ρ_1	ρ_2
ρ_3	ρ_2	ρ_1	1	ρ_1
ρ_4	ρ_3	ρ_2	ρ_1	1

Figure A6.1 The correlation matrix ρ_{ij} for the example $n = 5$. If all elements are summed by adding diagonally, one obtains $5 + 2(4\rho_1 + 3\rho_2 + 2\rho_3 + \rho_4)$.

For series that are much longer than the correlation length ($k \ll n$), (A6.5) can be reduced to

$$n_c = 1 + 2\sum_{k=1}^{\infty} \rho_k. \tag{A6.7}$$

Equations (A6.4) and (A6.5) follow simply from counting the number of occurrences in the double sum of (A6.4). Figure A6.1 elucidates how (A6.5) is obtained by summing all matrix elements.

Instead of (A6.3) we now obtain as a result:

$$E[\langle(\Delta x)^2\rangle] = \sigma^2\left(1 - \frac{n_c}{n}\right). \tag{A6.8}$$

The effect of correlation in the data series on the estimate of σ is not very large and can generally be neglected. However, the effect on the estimate for the standard inaccuracy of the average is quite large and not negligible. This is treated in Appendix A7.

A7 Standard deviation of the mean

Why is the variance of the mean $\langle x \rangle$ of n independent data equal to the variance of x itself divided by n?

We investigate the following quantity:

$$\text{var}(\langle x \rangle) = E[(\langle x \rangle - \mu)^2] = \frac{1}{n^2} E\left[\left\{\sum_i (x_i - \mu)\right\}^2\right] \quad (A7.1)$$

$$= \frac{1}{n^2} \sum_i \sum_j E[(x_i - \mu)(x_j - \mu)]. \quad (A7.2)$$

For uncorrelated data $E[(x_i - \mu)(x_j - \mu)] = \sigma^2 \delta_{ij}$. Therefore

$$\text{var}(\langle x \rangle) = \sigma^2/n \quad (A7.3)$$

and

$$\sigma_{\langle x \rangle} = \frac{1}{\sqrt{n}} \sigma. \quad (A7.4)$$

How is this result influenced when the data are correlated?

For this we need to work out the double sum in (A7.2). We have already done that in Appendix A6, see (A6.4), for the case of an ordered sequence in which correlations depend only on the distance $k = |j - i|$. It was found that

$$\text{var}(\langle x \rangle) = \sigma^2 \frac{n_c}{n}, \quad (A7.5)$$

where n_c is the correlation length, defined in Appendix A6 in (A6.7). So you see that correlations in the data tend to increase the uncertainty of the mean. It is as if the effective number of data points is less than the number you actually have. In order to make a reliable estimate of the uncertainty, you need to know the correlation length n_c, or deduce n_c from the data. This is in general not a simple task because the correlation between data points is difficult to evaluate, especially for large intervals. The correlation length is an integral over the correlation function, which is notoriously difficult to determine from noisy data.[1]

A practical alternative to the summing of correlation coefficients is the *block average* procedure:[2] group blocks of sequential data together and consider the average of each block as a new data point. If most of the sequential correlation is located within a block, the block averages are mutually almost uncorrelated and can be treated by standard methods. For example, if you have 1000 data points and you expect the correlation to stretch over some 10 or 20 points, then choose 10 blocks of 100 points each. Much better is to vary the block length and check if the results have a reliable limit. This "block average" procedure is not exact because there is always some correlation left between successive blocks, but it is very practical.

Example

Time series generated by Monte Carlo or molecular dynamics simulations often contain significant sequential correlations that complicate the determination of the inaccuracies of averages. In a dynamic simulation of a molecular system a time series of 20 000 data points (t, T) is generated with the "temperature" T (derived from the total kinetic energy) at times t in steps of 0.009 ps. Applying the rules for uncorrelated samples, the average temperature appears to be 309.967 ± 0.022 K. This inaccuracy is likely to be far too low in view of the expected sequential correlation of the data points T. What is the true standard inaccuracy? Figure A7.1 plots the first 500 points versus time: it is apparent that correlation persists over times of the order of picoseconds and includes some oscillatory behavior. Figure A7.2 plots the standard inaccuracy estimated from a series of block averages, with block sizes varying between 1 and 400 points, or 0.009 to 3.6 ps. A plateau of 0.11 K is reached after about 2 ps. This plateau value is 5 times larger than the "uncorrelated value," indicating that the statistical correlation length n_c is some 25 time steps or 0.22 ps. The block length must be taken several times larger than the correlation length for the blocks to be statistically independent. The final result for the average temperature is 309.97 ± 0.11 K.

---

[1] A discussion of alternative methods to determine the accuracy of the mean of correlated data can be found in Hess (2002), see reference list on page 124.

[2] The method is similar, but not identical to the "jackknife procedure," which estimates the mean and variance by averaging over data sets from which subgroups of data have been omitted. See Wolter (2007), Chapter 4, in the reference list on page 124.

Simulated temperature

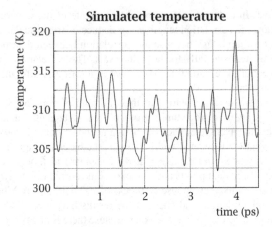

Figure A7.1 The first 500 points of a 20 000 point data set with temperatures derived from the kinetic energy as a function of time of a molecular dynamics simulation of a molecular system. The time interval between points is 0.009 ps.

Block-averaged s.d.

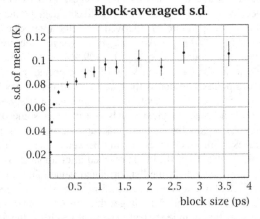

Figure A7.2 Estimates of the inaccuracy (standard deviation) in the mean using block averages of 20 000 data points with temperature data (from a molecular dynamics simulation) as function of time. The block averages are assumed to be uncorrelated. The block size varies from 1 point (0.009 ps) to 400 points (3.6 ps). The error bars indicate the uncertainties in the s.d. based on the limited number n_b of block averages, which amounts to a relative error of $1/\sqrt{2(n_b - 1)}$.

> **Python code** 7.1 on page 195 shows how the standard inaccuracy of averages can be estimated from a set of block averages.

How accurate is the estimated standard deviation?

Because the variance of a distribution is estimated from the sum of squared deviations from the average (divided by $n - 1$), the statistics of the variance satisfies the statistics of a sum of squares of random samples. For normally distributed samples, this sum follows the "chi-squared distribution" (see Section 7.4 and data sheet CHI-SQUARED DISTRIBUTION on page 199). A chi-squared distribution has a mean ν and a variance 2ν; hence it has a relative s.d. of $\sqrt{2/\nu}$, where ν is the number of degrees of freedom: $\nu = n - 1$. So the relative s.d. of the variance is $\sqrt{2/(n-1)}$ and the relative s.d. of the standard deviation itself is $1/\sqrt{2(n-1)}$. This result is valid for normally distributed independent samples. Any sequential correlation will increase the inaccuracy.

A8 Weight factors when variances are not equal

What is the "best" determination of the mean of a number of data x_i with the same expectations μ but with unequal standard deviations σ_i?

The answer is: take a *weighted average*:

$$\langle x \rangle = \frac{1}{w} \sum_{i=1}^{n} w_i x_i; \quad w = \sum_{i=1}^{n} w_i. \tag{A8.1}$$

But the question remains: how should you choose the w's? What criterion for the "best" choice is valid here? The criterion that the estimate of the mean should be *unbiased*, i.e., that the expectation of the mean should be equal to μ, is not useful because this is true for any choice of weight factors. The next obvious criterion is the *minimal variance estimate*: the most sharp and hence most accurate value. So let us determine w_i such that

$$E[(\langle x \rangle - \mu)^2] = E[\langle x - \mu \rangle^2] \, minimal, \tag{A8.2}$$

or

$$E[\langle x - \mu \rangle^2] = E\left[\frac{1}{w^2} \left(\sum_i w_i(x_i - \mu) \right)^2 \right]$$

$$= \frac{1}{w^2} \sum_{i,j} w_i w_j E[(x_i - \mu)(x_j - \mu)]$$

$$= \frac{1}{w^2} \sum_{i,j} w_i w_j \, \mathrm{cov} \, (x_i, x_j) = minimal. \tag{A8.3}$$

Now assume that x_i and x_j are uncorrelated, meaning that in the summation only $j = i$ survives. So we search for the minimum of the quantity $\sum_i w_i^2 \sigma_i^2$ under the condition that $\sum_i w_i$ remains constant. The standard way to solve such an *optimization with boundary condition* problem is Lagrange's

158

method of undetermined multipliers. In this method the boundary condition ($\sum_i w_i$ is constant) is multiplied by an as-yet-undetermined multiplier λ and then added to the function that is to be minimized. The partial derivatives of this total function with respect to each of the variables is then set to zero. The solution of the obtained set of equations still contains the undetermined multiplier, but the latter follows from the boundary condition. This is the way it goes:

$$\frac{\partial}{\partial w_i}\left(\sum_j w_j^2 \sigma_j^2 + \lambda \sum_j w_j\right) = 2w_i \sigma_i^2 + \lambda = 0. \qquad \text{(A8.4)}$$

Therefore

$$w_i \propto \frac{1}{\sigma_i^2}. \qquad \text{(A8.5)}$$

The conclusion must be that the weight of each data point must be proportional to the inverse variance of that point. This is valid when the deviations of the data points are uncorrelated.

The same conclusion can be reached if it is *assumed* that the distribution of deviations is normal. However, the requirement of minimal variance is much more general and the result applies to any distribution function with finite variance.

How large is the variance in $\langle x \rangle$?

For this the expectation of $(\langle x \rangle - \mu)^2$ must be computed:

$$\sigma_{\langle x \rangle}^2 = E[(\langle x \rangle - \mu)^2] = \frac{1}{w^2}\sum_i w_i^2 \sigma_i^2.$$

Here use is made of the fact that x_i and x_j are uncorrelated. For w_i we choose $w_i = 1/\sigma_i^2$ and it follows that

$$\sigma_{\langle x \rangle}^2 = \frac{1}{w^2}\sum_i \frac{1}{\sigma_i^2} = \left(\sum_i \frac{1}{\sigma_i^2}\right)^{-1}. \qquad \text{(A8.6)}$$

A9 Least-squares fitting

In this appendix *matrix notation* is used. A bold lower case letter is a column matrix (which is an $n \times 1$ matrix representing a vector); a bold capital letter is a matrix. A matrix product $C = AB$ is defined by $C_{ij} = \sum_k A_{ik}B_{kj}$. The transpose A^T of A is defined by $(A^T)_{ij} = A_{ji}$. The trace $\text{Tr}(A)$ is the sum of diagonal elements of A. The inverse A^{-1} fulfills $A^{-1}A = AA^{-1} = 1$ (unit matrix). Recall that $(AB)^T = B^TA^T$ and $(AB)^{-1} = B^{-1}A^{-1}$. The trace of a matrix product is invariant for cyclic permutation of its terms: $\text{Tr}(ABC) = \text{Tr}(CAB)$. Note that for a column matrix (vector) a the product a^Ta is a scalar equal to $\sum_i a_i^2$, while aa^T is a square matrix with elements a_ia_j.

A9.1 How do you find the best parameters a and b in y ≈ ax + b?

In order to find the values of a and b in the function $f(x) = ax + b$, such that

$$S = \sum_{i=1}^n w_i(y_i - f_i)^2 = \sum_{i=1}^n w_i(y_i - ax_i - b)^2 \text{minimal,}$$

you simply solve for zero derivatives of S/w ($w = \sum_i w_i$) with respect to a and b:

$$\frac{1}{w}\frac{\partial S}{\partial a} = -\frac{2}{w}\sum_{i=1}^n w_i x_i(y_i - ax_i - b) = 0$$

$$\frac{1}{w}\frac{\partial S}{\partial b} = -\frac{2}{w}\sum_{i=1}^n w_i(y_i - ax_i - b) = 0.$$

From the second equation it follows that $b = \langle y \rangle - a\langle x \rangle$. Substitution of b in the first equation yields the solution for a, see (7.13). The averages are weighted averages such as

$$\langle y \rangle = \frac{1}{w} \sum_{i=1}^{n} w_i y_i.$$

A9.2 General linear regression

In general a set of equations linear in m parameters $\theta_k, k = 1, \ldots, m$ can be written as

$$f_i(\theta_1, \theta_2, \ldots, \theta_m) = \sum_{k=1}^{m} A_{ik}\theta_k, \quad \text{of } f(\theta) = A\theta. \qquad (A9.1)$$

Suppose that the "true" values y_i are given by

$$y = A\theta_m + \epsilon, \qquad (A9.2)$$

where θ_m are the "true" *model values* of the parameters and ϵ the added stochastic variable or "noise" with properties

$$E[\epsilon] = 0 \qquad (A9.3)$$

$$E[\epsilon\epsilon^{\mathrm{T}}] = \Sigma. \qquad (A9.4)$$

Here Σ is the *covariance matrix* of "errors" ϵ in the measured values y. This is a very general assumption allowing correlation between the data points. If Σ is diagonal, the data are not correlated.

The chi-squared sum can now be written as

$$\chi^2 = (y - A\theta)^{\mathrm{T}} \Sigma^{-1} (y - A\theta). \qquad (A9.5)$$

Now the case is quite common that Σ is not accurately known, and you only know something about the relative size and the mutual correlation of the data. So assume that – on the basis of your limited knowledge of the uncertainties – you can assign a *weight matrix* W that is proportional to the inverse of the covariance matrix of the random errors in the measured values:

$$W = c\Sigma^{-1}. \qquad (A9.6)$$

For the moment the constant c is unknown, but – as we shall see below – under certain conditions c is derivable from the data themselves. Without correlations between the data points both Σ and W are diagonal, with σ_i^2, resp. $c\sigma_i^{-2}$, as diagonal elements.

We can now construct the SSQ: the Sum of (weighted) SQuare deviations S:

$$S = (y - A\theta)^{\mathrm{T}} W (y - A\theta) = c\chi^2. \qquad (A9.7)$$

The derivatives of S with respect to the parameters yields the following vector:

$$\frac{\partial S}{\partial \theta} = -2A^T W(y - A\theta) = 0. \tag{A9.8}$$

The least-squares solution for θ, indicated by $\hat{\theta}$, is the solution of the set of equations

$$A^T W A \theta = A^T W y. \tag{A9.9}$$

Thus the final solution for the best estimate of θ is

$$\hat{\theta} = (A^T W A)^{-1} A^T W y. \tag{A9.10}$$

This equation solves any linear least-squares fit, including multiple explanatory variables and including any known correlations between data points. Note that the exact values of the individual inaccuracies are not needed to determine the minimum: if all values of W are multiplied by a constant, the solution $\hat{\theta}$ does not change.

The least-squares solution $\hat{\theta}$ is an *unbiased* estimate of θ, meaning that the expectation of the estimate equals the true value:

$$E[\hat{\theta}] = (A^T W A)^{-1} A^T W E[y] = \theta_m, \tag{A9.11}$$

because, according to (A9.2) and (A9.3), $E[y] = A\theta_m$.

A9.3 SSQ as a function of the parameters

The expression (A9.7) for $S(\theta)$ can be written as a quadratic function of the parameters. After relating S to χ^2, we find the likelihood $\exp[-\frac{1}{2}\chi^2]$ – see (7.4) on page 86 – as a quadratic function of the parameters and from that we can estimate the variances and covariances of the parameters.

Defining the *deviations* from the best estimates of the parameters:

$$\Delta\theta \stackrel{\text{def}}{=} \theta - \hat{\theta}, \tag{A9.12}$$

and the minimum of S:

$$S_0 = (y - A\hat{\theta})^T W(y - A\hat{\theta}), \tag{A9.13}$$

and inserting (A9.10) and (A9.12) into (A9.13), we find

$$S(\theta) = S_0 + \Delta\theta^T A^T W A \Delta\theta. \tag{A9.14}$$

Here the gradient A9.8 has been used. So you see that S is a parabolic function in $\boldsymbol{\Delta\theta}$.

Since the likelihood depends on $\chi^2 = S/c$, we need to estimate c. This is straightforward as the expectation for χ_0^2 equals the number of degrees of freedom $n - m$:

$$\hat{\chi}_0^2 = \frac{S_0}{c} = n - m. \tag{A9.15}$$

Hence $c = S/(n - m)$ and

$$\hat{\chi}^2(\boldsymbol{\theta}) = n - m + \frac{n - m}{S_0} \boldsymbol{\Delta\theta}^{\mathrm{T}} \boldsymbol{A}^{\mathrm{T}} \boldsymbol{W} \boldsymbol{A} \boldsymbol{\Delta\theta} \tag{A9.16}$$

$$= n - m + \boldsymbol{\Delta\theta}^{\mathrm{T}} \boldsymbol{B} \boldsymbol{\Delta\theta}, \tag{A9.17}$$

where

$$\boldsymbol{B} \stackrel{\text{def}}{=} \frac{n - m}{S_0} \boldsymbol{A}^{\mathrm{T}} \boldsymbol{W} \boldsymbol{A}. \tag{A9.18}$$

From (A9.17) you see that the matrix of second derivatives of $\chi^2(\boldsymbol{\theta})$ is given by $2\boldsymbol{B}$.

The likelihood P (proportional to $\exp[-\frac{1}{2}\chi^2]$) is of the form:

$$P \propto \exp\left[-\frac{1}{2}\boldsymbol{\Delta\theta}^{\mathrm{T}}\boldsymbol{B}\boldsymbol{\Delta\theta}\right]. \tag{A9.19}$$

In case you have reliable knowledge on the uncertainties $\boldsymbol{\Sigma}$, so that you can take the weight matrix *exactly* equal to $\boldsymbol{\Sigma}^{-1}$, the likelihood is

$$P \propto \exp\left[-\frac{1}{2}\boldsymbol{\Delta\theta}^{\mathrm{T}}\boldsymbol{A}^{\mathrm{T}}\boldsymbol{\Sigma}^{-1}\boldsymbol{A}\boldsymbol{\Delta\theta}\right]. \tag{A9.20}$$

Both forms are multivariate normal distributions. With this knowledge we can derive the (co)variances of the parameters.

A9.4 Covariances of the parameters

A multivariate normal distribution (see data sheet NORMAL DISTRIBUTION on page 205) has the form

$$P \propto \exp\left[-\frac{1}{2}\boldsymbol{\Delta\theta}^{\mathrm{T}}\boldsymbol{C}^{-1}\boldsymbol{\Delta\theta}\right], \tag{A9.21}$$

where $\boldsymbol{C}$ is the covariance matrix:

$$\boldsymbol{C} = E[(\boldsymbol{\Delta\theta})(\boldsymbol{\Delta\theta})^{\mathrm{T}}] \tag{A9.22}$$

$$\boldsymbol{C}_{kl} = \operatorname{cov}(\Delta\theta_k, \Delta\theta_l). \tag{A9.23}$$

Comparing this to the likelihood expressions (A9.19) and (A9.20), the expressions for the covariance matrix are found. For the common case that S_0 is used to estimate χ^2:

$$C = B^{-1}; \quad B \text{ defined in (A9.18)} \tag{A9.24}$$

and for the case that uncertainties Σ are accurately known:

$$C' = \left(A^{\mathsf{T}}\Sigma^{-1}A\right)^{-1}. \tag{A9.25}$$

These are our main results. Practical equations are simplifications of (A9.24) and (A9.25).

In order to simplify the presentation, consider the case that there is no correlation between data points, and their variances are σ_i^2 so that $\Sigma = \mathrm{diag}\,(\sigma_i^2)$ and $W = c\,\mathrm{diag}\,(\sigma_i^{-2})$. Then the covariance matrix (A9.24) simplifies to

$$C = B^{-1}; \quad B_{kl} = \frac{n-m}{S_0} \sum_i w_i A_{ik} A_{il} \tag{A9.26}$$

and (A9.25) simplifies to

$$C' = B'^{-1}; \quad B'_{kl} = \sum_i \sigma_i^{-2} A_{ik} A_{il}. \tag{A9.27}$$

The equations for the parameter (co)variances for linear regression of $f(x) = ax + b$, given in Chapter 7 on page 89 in (7.18), (7.19) and (7.20), are easily recovered from these equations. For $\theta_1 = a$ and $\theta_2 = b$, the $n \times 2$ matrix A is given by

$$A_{i1} = x_i; \quad A_{i2} = 1. \tag{A9.28}$$

For example, the element B_{11} of the 2×2 matrix B (A9.26) can be written as

$$B_{11} = \frac{n-m}{S_0} \sum w_i x_i^2 = \frac{n-m}{S_0} \frac{1}{w} \langle x^2 \rangle, \tag{A9.29}$$

where w is the total sum of w_i. The rest of the derivation is straightforward and left to the reader.

Why is the s.d. of a parameter given by the projection of the ellipsoid $\Delta\chi^2 = 1$?

The condition $\Delta\chi^2 = 1$ describes a surface (an ellipsoid) in the m-dimensional parameter space. In Fig. 7.5 on page 103 tangents to the ellipse $\Delta\chi^2 = 1$ indicate that the *projection* of this figure on one of the axes (e.g. θ_1) occurs within the limits $\hat{\theta}_1 \pm \sigma_1$. The tangent touches the ellipse in a point where χ^2 is minimal with respect to all *other* parameters $\theta_2, \ldots, \theta_m$, i.e., where the gradient of χ^2 points in the direction of θ_1:

$$\mathbf{grad}\,\chi^2 = (a, 0, \ldots, 0)^{\mathrm{T}},$$

where a is a constant resulting from $\Delta\chi^2 = \Delta\boldsymbol{\theta}^{\mathrm{T}}\boldsymbol{B}\Delta\boldsymbol{\theta} = 1$: because[1] $\mathbf{grad}\,\chi^2 = 2\boldsymbol{B}\Delta\boldsymbol{\theta}$,

$$\Delta\boldsymbol{\theta}^{\mathrm{T}}\frac{1}{2}(a, 0, \ldots, 0)^{\mathrm{T}} = \frac{1}{2}a\Delta\theta_1 = 1.$$

Hence

$$\boldsymbol{B}\Delta\boldsymbol{\theta} = \frac{1}{2}(2/\Delta\theta_1, 0, \ldots, 0)^{\mathrm{T}}$$

and

$$\Delta\boldsymbol{\theta} = \boldsymbol{C}(1/\Delta\theta_1, 0, \ldots, 0)^{\mathrm{T}} \text{ or } \Delta\theta_1 = \pm\sqrt{C_{11}} = \pm\sigma_1. \tag{A9.30}$$

This is what we wished to prove.[2]

Nonlinear least-squares fit

When the functions $f_i(\theta_1, \ldots, \theta_m)$ are not linear in all parameters, but $S = (\boldsymbol{y} - \boldsymbol{f})^{\mathrm{T}}\boldsymbol{W}(\boldsymbol{y} - \boldsymbol{f})$ does have a minimum $S_0 = S(\hat{\boldsymbol{\theta}})$, then $S(\boldsymbol{\theta})$ can be expanded around that minimum in a Taylor series with zero linear term, just as in (A9.14) in the linear case. In terms of the expectation of χ^2 (which equals $n - m$ at the minimum):

$$\hat{\chi}^2(\boldsymbol{\theta}) = \frac{n - m}{S_0}S(\boldsymbol{\theta}) = n - m + \Delta\boldsymbol{\theta}^{\mathrm{T}}\boldsymbol{B}\Delta\boldsymbol{\theta} + \cdots, \tag{A9.31}$$

After a redefinition of the matrix $\boldsymbol{A}$:

$$A_{ik} = \left(\frac{\partial f_i}{\partial\theta_k}\right)_{\hat{\theta}}, \tag{A9.32}$$

all equations for the parameters and their (co)variances remain approximately valid. The inverse of $\boldsymbol{B} = \frac{n-m}{S_0}\boldsymbol{A}^{\mathrm{T}}\boldsymbol{W}\boldsymbol{A}$ is still (but approximately) equal to the covariance matrix of the parameters. See Press *et al.* (1992)[3] for

[1] The gradient of a quadratic form $\frac{1}{2}\boldsymbol{x}^{\mathrm{T}}\boldsymbol{G}\boldsymbol{x}$, with $\boldsymbol{G}$ symmetric, equals $\boldsymbol{G}\boldsymbol{x}$.
[2] The proof can be found in Press *et al.* (1992), see the reference section on page 124.
[3] See reference list on page 124.

a discussion on this point. For uncorrelated data, Equation (A9.26) on page 164 is still valid:

$$B_{kl} = \frac{n-m}{S_0} \sum_{i=1}^{n} w_i \frac{\partial f_i}{\partial \theta_k} \frac{\partial f_i}{\partial \theta_l}. \tag{A9.33}$$

The covariance matrix is approximately equal to the inverse of B.

For functions that are not linear in the parameters, the likelihood function is only approximately equal to a multivariate normal distribution. Especially the tails of the distribution may differ and the blind derivation of confidence limits based on normal distributions may be erroneous in the tail regions of the distribution. More accurate estimations can be done using the likelihood function

$$p(\theta) \propto \exp\left[-\frac{1}{2}\chi^2(\theta)\right]. \tag{A9.34}$$

As the practical implications are of little importance, we don't pursue this point here any further.

PART III

Python codes

This appendix contains programs, functions or code fragments written in Python. Each code is referred to in the text; the page where the reference is made is given in the header.

First some general instructions are given on how to work with these codes. Python is a general-purpose interpretative language, for which interpreters are available for most platforms, including Windows. Python is in the public domain and interpreters are freely available.[1] Most applications in this book use a powerful numerical array extension *NumPy*, which also provides basic tools in linear algebra, Fourier transforms and random numbers.[2] Although Python version 3 is available, at the time of writing NumPy requires Python version 2, the latest being 2.6. In addition, applications may require the scientific tools library *SciPy*, which relies on *NumPy*.[3] Importing *SciPy* automatically implies the import of *NumPy*.

Users are advised first to download Python 2.6, then the most recent stable version of *NumPy*, and then *SciPy*. Further instructions for Windows users can be found at www.hjcb.nl/python.

There are several options to produce plots, for example *Gnuplot.py*,[4] based on the *gnuplot* package[5] or *rpy*[6] based on the statistical package "R."[7] But there are many more.[8] Since the user may find it difficult to make a choice, we have added yet another, but very simple to use, plotting module called `plotsvg.py`. It can be downloaded from the author's website.[9] Its plotting routines produce SVG output files (Scalable Vector Graphics, a W3C standard) that can be viewed by an SVG-enabled browser. Among others, the Firefox, Opera and Google Chrome browsers (but not Internet Explorer) have native SVG support. While customized plots are possible, automatic plots of functions, points and cumulative distributions can be simply made. For example, the following code produces automatic display of the cumulative distribution of 200 normally distributed random numbers on a probability scale (on which normal distributions should give a straight line):

Python code 0.1 *Demo plotsvg*

---

[1] www.python.org.
[2] www.scipy.org/numpy.
[3] www.scipy.org/SciPy.
[4] http://gnuplot-py.sourceforge.net/.
[5] www.gnuplot.info/.
[6] http://rpy.sourceforge.net/.
[7] www.r-project.org/.
[8] See http://wiki.python.org/moin/NumericAndScientific/Plotting.
[9] www.hjcb.nl/python.

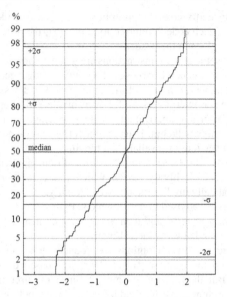

Figure P.1 Plotting demo: a cumulative plot on a probability scale of 200 random samples from a normal distribution.

```
from scipy import *
from plotsvg import *
r=randn(200)
autoplotc(r,yscale='prob')
```

with Fig. P.1 as a result.

Comments:

The module `plotsvg.py` defines a class `Figure()` with methods `frame()` to define a frame with titles allowing for logarithmic and probability scales, `plotp()` to plot a series of points with or without connecting lines and error bars, `plotc()` to plot cumulative distributions, `plotf()` to plot functions and a number of utilities like `addtext()` and `addobject()`. There are also stand-alone programs like `autoplotp()` for quick graphs.

Another module that can be downloaded from the author's website is *physcon.py*. This module contains most of the fundamental physical constants as SI values in the form of a dictionary. In addition the following symbols are defined as the SI value (float): `alpha`, `a_0`, `c`, `e`, `eps_0`, `F`, `G`, `g_e`, `g_p`, `gamma_p`, `h`, `hbar`, `k_B m_d`, `m_e`, `m_n`, `m_p`, `mu_B`, `mu_e`, `mu_N`, `mu_p`, `mu_0`, `N_A`, `R`, `sigma`, `u`.

Python code 0.2 *Demo physcon*

```
import physcon as pc
pc.help()
```

This will list available functions, variables and keys

```
pc.descr('avogadro')
```

This will describe avogadro: name, symbol, value, standard error, relative s.d., unit, data source

```
N=pc.N_A
```

This will assign the value $6.02214179e + 023$ to N

Python code 2.1 (page 6) *Generate and plot Fig. 2.2*

```
from scipy import *
x = 8.5 + randn(30)
xr = x.sort().round(2)
from plotsvg import *
autoplotc(xr,title='Cumulative distribution')
autoplotc(xr,title='Cumulative distribution',\
          yscale='prob')
```

Python code 2.2 (page 8) *Generate histogram of Fig. 2.3*

```
from plotsvg import *
hisx = [6.5,7.5,8.5,9.5,10.5,11.5]
hisy = [1,7,8,10,2,2]
f = Figure()
f.frame([6,12],title='Histogram')
f.plotp([hisx,hisy],symbol='halfbar',\
    symbolfill=Darkgrey,symbolstroke=Black)
f.show()
```

Python code 2.3 (page 8) *Some array methods and functions*

```
from scipy import *
n=alen(x)      # assigns length of array x to n
m=x.mean()     # assigns mean of x to m
msd=x.var()    # assigns mean squared deviation
               # of x to msd
```

```
rmsd=x.std()   # assigns root mean squared deviation
               # of x to rmsd
```

Python code 2.4 (page 8) *Generate percentiles of a given data set*

```
from scipy import *
from scipy import stats
def percentiles(x, per=[1,5,10,25,50,75,90,95,99]):
# x = 1D-array
# per = list of percentages
    scores=zeros(len(per),dtype=float)
    i=0
    for p in per:
        scores[i]=stats.scoreatpercentile(x,p)
        i++
    return scores
```

Comments:
The scipy.stats function `scoreatpercentile(x,p)` gives the p-th percentile, i.e. the value $\geq p\%$ of the data and $\leq (100 - p)\%$ of the data. If that is not a single value, linear interpolation is used.

Python code 2.5 (page 16) *Plot on a logarithmic scale*

```
from plotsvg import *
time=[20.,40.,60.,80.,100.,120.,140.,160.,180.]
conc=[75.,43.,26.,16.,10.,5.,3.5,1.8,1.6]
err=[4.,3.,3.,3.,2.,2.,1.,1.,1.]
f=Figure()
f.frame([[0,200],[1,100]],xlabel='time <i>t</>/s',\
    ylabel='concentration <i>c</i>/mmol L<sup>-1\
    </sup>', yscale='log')
f.plotp([time,conc],ybars=err)
f.show()
```

Comments:
This code produces Fig. 2.7 with the module `plotsvg` available from the author's web site www.hjcb.nl/python/. An SVG file is produced and displayed by a suitable browser (Firefox, Opera, Google Chrome, but not Internet Explorer).

Python code 3.1 (page 25) *Monte Carlo generation of equilibrium constant*

```python
from scipy import *
from plotsvg import *
def Keq(a,b,V1,V2,x):                    # define equilibrium
                                         # constant
    V=V1+V2
    K=x/((a/V-x)*(b/V-x))*1000.          # convert to L/mol
    return K
n=1000                                   # set number of
                                         # samples
a0=5.0; a=a0+randn(n)*0.2                # mmol
b0=10.0; b=b0+randn(n)*0.2              # mmol
V10=0.1; V1=V10+randn(n)*0.001          # L
V20=0.1; V2=V20+randn(n)*0.001          # L
x0=5.0; x=x0+randn(n)*0.35              # mmol/L
K=Keq(a,b,V1,V2,x)                      # L/mol (array of
                                         # K-values)
K0=Keq(a0,b0,V10,V20,x0)                # L/mol (K at
                                         # central values)
print 'K from values without noise = %g' % (K0)
print 'number of samples = %d' % (n)
print 'average and std of K = %g +/- %g' %\
        (K.mean(), K.std())
```

Generate Figure 3.1:

```python
f=Figure()
f.size=[5500,6400]
f.frame([[4.,7.5],[0,100]],title='Equilibrium\
        constant', yscale='prob',\
        xlabel='<i>K</i><sub>eq</sub>/L mol<sup>-1\
        </sup>', ylabel='cumulative probability\
        distribution')
f.plotc(K)
f.show()
```

Python code 4.1 (page 36) *Generate binomial functions for Figures 4.1, 4.2 and 4.3*

```python
from scipy import *
from scipy import stats
```

Probability of finding k "heads" in 10 coin tossings:

```
def fun1(k): return stats.binom.pmf(k,10,0.5)
```

Probability of finding k "6"'s in 60 dice throws:

```
def fun2(k): return stats.binom.pmf(k,60,1./6.)
```

Probability of exceeding k correct guesses in 25 Zener cards:

```
def fun3(k): return stats.binom.sf(k,25,0.2)
```

Generate Fig. 4.1:

```
from plotsvg import *
x1=arange(11); y1=fun1(x1)
f=Figure()
f.frame([[-1,11],[-0.02,0.27]],\
    title="Binomial 10 coin tosses",\
    xlabel="nr of heads", ylabel="probability")
f.plotp([x1,y1], symbol='halfbar',\
    symbolstroke=Black, symbolfill=Darkgrey)
f.show()
```

Generate Fig. 4.2:

```
x2=arange(27); y2=fun2(x2)
f=Figure()
f.frame([[-1,26],[-0.01,0.15]],\
    title="Binomial 60 dice throws",\
    xlabel="nr of 6's", ylabel="probability")
f.plotp([x2,y2],symbol='halfbar',\
    symbolstroke=Black,symbolfill=Darkgrey)
f.show()
```

Generate Fig. 4.3:

```
x3=arange(16); y3=fun3(x3)
f=Figure()
f.frame([[0,12],[0,1]],\
    title="Binomial 25 Zener cards",\
    xlabel="nr correct", ylabel="survival\
    (1 - c.d.f.)")
f.plotp([x3,y3],symbol='dot',lines=Black)
f.show()
```

Python code 4.2 (page 47) *Generate Weibull distribution functions*

```
from scipy import stats
pdf=stats.weibull_min.pdf
cdf=stats.weibull_min.cdf
def f1(t):
    if (t<0.001):
        return None
    else: return pdf(t,0.5)
def g1(t): return cdf(t,0.5)
def f2(t): return pdf(t,1.)
def g2(t): return cdf(t,1.)
```

Comments:
The scipy module stats contains a large number of distribution functions. The pdf for negative c is infinity for $t = 0$, which should be excluded. The pdf's f_1, f_2 and the cdf's g_1, g_2 are suitable for plotting.

Python code 5.1 (page 67) *The bootstrap method: Generate averages from random samples*

```
def bootstrap(x,n,dof=0):
# x = 1D-array of input samples
# n = nr of averages generated
# dof = nr of degrees of freedom.
# If not specified, dof=len(x)
# returns 1D-array of averages
    from scipy import stats
    nx=len(x)
    if (dof==0): nu=nx
    else: nu=dof
    result=zeros(n,dtype=float)
    for i in range(n):
        index=stats.randint.rvs(0,nx,size=nu)
        result[i]=x[index].mean()
    return result
```

Comments:
The randint.rvs(min,max,size=n) function of the scipy package stats produces an array of n random integers $>=$ min and $<$ max.

x[index] produces an array containing the values of x[i], where i are all values of the integer array index.

If dof is not specified, the averages are taken over as many items as there are in the input array x; this yields the biased bootstrap distribution. The unbiased distribution can be approximated by setting dof equal to the length of x minus 1.

> **Python code 6.2** (page 67) **Report**: *a program that analyzes a set of independent data.*

```
from scipy import *
from plotsvg import *
def report(data,figures=True):
    '''
Function. Reports statistics on single uncorrelated
----------------------------------------------------
data series.
------------
arguments:
    data:          list or array [y] or [x,y] or [x,y,
                      sig] of data; if [y] then x=arange
                      (len(y))
                      sig = sd of y-value; if given,
                      chisq test reported, if sig not
                      given, equal weights are assumed
    figures=True if True, figures are produced and
                      displayed
returns:           [[mean,sdmean,var,sd],[a,siga,b,
                      sigb]] (fit ax+b)
Remarks:
    report of properties (average, msd, rmsd) and of
    estimates (mean, variance, sd, skewness, excess)
    and their accuracies is printed (skewness and
    excess only if relevant). Figures produced:
      figdata.svg: data points with error bars and
      linear fit;
      figcum.svg: cumulative plot on probability
      scale.
    Outliers are identified. If s.d. sig are given, a
    chi-squared analysis is produced. A linear
    regression  drift analysis is done.
    '''
    import os
    from scipy import stats
    # unify data structure:
    data=array(data)
    dimension=array(data).ndim
    weights=False
    if (dimension==1):
        n=len(data)
        xy=array([arange(n),data])
```

```python
    elif (dimension==2):
        n=len(data[0])
        if (len(data)==2):
            xy=array(data)
        elif (len(data)==3):
            xy=array(data[:2])
            weights=True
            w=1./array(data[2])**2
        else:
            print 'ERROR: wrong data length'
            print 'report aborted'
            return 0
    else:
        print 'ERROR: wrong data dimension'
        print 'report aborted'
        return 0
    # compute properties
    if weights:
        wtot=w.sum()
        xav=(xy[0]*w).sum()/wtot
        yav=(xy[1]*w).sum()/wtot
    else:
        wtot=float(n)
        xav=xy[0].mean()
        yav=xy[1].mean()
    xdif=xy[0]-xav
    ydif=xy[1]-yav
    if weights:
        ssq=(w*ydif**2).sum()
    else:
        ssq=(ydif**2).sum()
    msd=ssq/wtot
    rmsd=sqrt(msd)
    var=msd*n/(n-1.)
    sd=sqrt(var)
    sdmean=sd/sqrt(n)
    ymin=xy[1].min()
    yminindex=xy[1].argmin()
    ymax=xy[1].max()
    ymaxindex=xy[1].argmax()
    # linear regression:
    if weights:
        xmsd=(w*xdif**2).sum()/wtot
        a=(xdif*ydif*w).sum()/wtot/xmsd
```

```python
        b=yav-a*xav
        S=(w*((xy[1]-a*xy[0]-b)**2)).sum()
        siga=sqrt(S/(wtot*(n-2.)*xmsd))
        sigb=siga*sqrt((w*(xy[0]**2))).sum()/wtot)
    else:
        xmsd=(xdif**2).mean()
        a=(xdif*ydif).mean()/xmsd
        b=yav-a*xav
        S=((xy[1]-a*xy[0]-b)**2).sum()
        siga=sqrt(S/(n*(n-2.)*xmsd))
        sigb=siga*sqrt((xy[0]**2).mean())
    # produce figures
    if figures:
        def fun0(x): return yav
        def fun1(x): return yav-sd
        def fun2(x): return yav+sd
        def fun3(x): return yav-2.*sd
        def fun4(x): return yav+2.*sd
        def fun5(x): return a*x+b
        f=Figure()
        f.frame([[xy[0,0],xy[0,-1]],[ymin,ymax]],\
                title='input data')
        f.plotf(fun0,color=Red)
        f.plotf(fun1,color=Red)
        f.plotf(fun2,color=Red)
        f.plotf(fun3,color=Red)
        f.plotf(fun4,color=Red)
        f.plotf(fun5,color=Green)
        if weights:
            f.plotp(xy,ybars=data[2],symbolfill=\
                    Blue, barcolor=Blue)
        else:
            f.plotp(xy,lines=Blue,symbol='')
        f.addtext([890,4140],\
            '<small>red lines: mean, &#177; &#963;,\
            &#177; 2&#963; </small>',fill=Red)
        f.addtext([4890,4140],\
            '<small>green line: linear\
            regression</small>', align='r',\
            fill=Green)
        f.show(filename='figdata.svg')
        os.startfile('figdata.svg')
        print 'figdata.svg is now displayed by\
            your browser'
```

```
            f=Figure()
            f.size=[5500,6400]
            f.frame([[(1.1*ymin-0.1*ymax),(-0.1*ymin\
                    +1.1*ymax)],\
                [0,100]], title='cum.distribution\
                of data', yscale='prob')
            f.plotc(xy[1])
            f.show(filename='figcum.svg')
            os.startfile('figcum.svg')
            print 'figcum.svg is now displayed by your\
                browser'
        print '\nStatistical report on uncorrelated\
            data series'
        print '\nProperties:'
        print 'nr of elem. = %5d' % n
        print 'average      = %10.6g' % (yav)
        print 'msd          = %10.6g' % (msd)
        print 'rmsd         = %10.6g' % (rmsd)
        print '\nEstimates'
        print 'mean         = %10.6g +/- %8.4g' % (yav,\
            sdmean)
        if weights: print '*)'
        print '\nvariance    = %10.6g +/- %8.4g' %\
            (var, var*sqrt(2./(n-1.)))
        print 'st. dev      = %10.6g +/- %8.4g' %\
            (sd, sd/sqrt(2.*(n-1.)))
        if weights:
            print '*) this standard uncertainty in the
                    mean is\
                    derived from the data variance'
            print "    derived from the supplied sigma's
                    it is", "%8.4g" % (wtot**(-0.5))
            print '  Choose the more reliable, or else\
                    the larger value.'
            print '   See also the chi-square analysis\
                    below.'
        # skewness and excess only if weights=False
        if not weights:
            if (n>=20):
                skew=(xy[1]**3).sum()/(n*var*sd)
                print 'skewness    = %10.6g +/- %8.4g'\
                    % (skew, sqrt(15./n))
            else: print 'skewness: insufficient\
            statistics'
```

```python
      if (n>=100):
          exc=(xy[1]**4).sum()/(n*var*var)-3.
          print 'excess        = %10.6g +/- %8.4g'\
                % (exc, sqrt(96./n))
      else: print 'excess: insufficient\
      statistics'
# outliers and their probabilities
ydevmax=(ymax-yav)/sd
ydevmin=(yav-ymin)/sd
Fmax=stats.norm.cdf(-ydevmax)
probmax=100.*(1.-(1.-Fmax)**n)
Fmin=stats.norm.cdf(-ydevmin)
probmin=100.*(1.-(1.-Fmin)**n)
print '\nPossible outliers:',
if ((probmax>5.) and (probmin>5.)):
    print '   (there are no significant\
    outliers with p<5%)'
else:
    print '(there are significant outliers\
    with p<5%)'
print 'largest element y[%d]=%10.6g deviates\
    +%5.2g stand.',\
      'dev. from mean'  % (ymaxindex,ymax,\
      ydevmax)
print 'prob. to obtain a higher value at least\
    once is',\
      '%4.3g %%' % (probmax)
print 'smallest element y[%d]=%10.6g deviates\
    -%5.2g stand.',\
      'dev from mean' % (yminindex,ymin,\
      ydevmin)
print 'prob. to obtain a lower value at least\
once is',\
      '%4.3g %%' % (probmin)
# chi-square analysis if weight=True:
if weights:
    nu=n-1
    F=stats.chi2.cdf(S,nu)
    print '\nChi-square analysis:'
    print 'chi^2 (sum of weighted square dev.)\
        = %10.6g' % (ssq)
    print 'cum. prob. for chi^2 = %5.3g %%' %\
        (100.*F)
    if (F<.1):
```

```
            print "chi^2 is low! Did you\
                overestimate the\
                supplied sigma's?"
            print 'Or did you fit the original\
                data too closely\
                with too many parameters?'
        elif (F>.9):
            print "chi^2 is high! Did you neglect\
                an error source\
                in the supplied sigma's?"
            print 'Or did the data result from a\
                bad fitting procedure?'
        else:
            print 'cum. probab. of chi^2 is\
                reasonable (between\
                10% and 90%).'
            print "The spread in the data agrees\
                with the supplied sigma's."
# Significance of drift
print '\nLinear regression: y=a*x+b'
print 'a=%10.6g +/- %10.6g; b=%10.6g +/-\
    %10.6g' % (a,siga,b,sigb)
Pdrift=2.*(1.-stats.norm.cdf(abs(a)/siga))
print '\nNormal test on significance of\
    slope a'
print 'Probability to obtain at least this\
    drift by random\
    fluctuation is %8.3g %%' %\
    (100.*Pdrift)
print '\nF-test on significance of linear\
    regression:'
print 'sum of square deviations reduced from\
    %7.5g to %7.5g'\
    % (ssq,S)
ypred=a*xy[0]+b
ypmean=ypred.mean()
if weights:
    SSR=(w*(ypred-ypmean)**2).sum()
else:
    SSR=((ypred-ypmean)**2).sum()
Fratio=SSR/(S/(n-1))
Fcum=stats.f.cdf(Fratio,1,n-1)
print 'The F-ratio SSR/(SSE/(n-1)) = %7.3g' %\
    (Fratio)
```

```
print 'The cum. prob. of the F-distribution is\
      %8.5g' % (Fcum)
print 'Probability to obtain this fit (or\
      better) by random',\
      'fluctuation is %8.3g %%' % (100.*\
      (1.-Fcum))
if ((Fcum>0.9) and (Pdrift<0.1)):
    print '\nThere is a significant drift (90%\
          conf. level)'
else:
    print '\nThere is no significant drift\
          (90% conf. level)'
print
return [[yav,sdmean,var,sd],[a,siga,b,sigb]]
```

Comments:

This program can be downloaded from www.hjcb.nl/python). Look for recent updates. Two plots are automatically generated and displayed by the standard browser. Make sure that the .svg mime type starts your SVG-enabled browser.

Python code 6.1 (page 81) *Fit a number of harmonics to data points*

```
from scipy import optimize
# data from compass corrections:
x=arange(0.,365.,15.)
y=array([-1.5,-0.5,0.,0.,0.,-0.5,-1.,-2.,-3.,-2.5,\
        -2.,-1.,0., 0.5,1.5,2.5,2.0,2.5,1.5,0.,\
        -0.5,-2.,-2.5,-2.,-1.5])
def fitfun(x,p):
    phi=x*pi/180.
    result=p[0]
    for i in range(1,5,1):
        result=result+p[2*i-1]*sin(i*phi)+p[2*i]\
        *cos(i*phi)
    return result                    # result is
                                     # array like x
def residuals(p): return y-fitfun(x,p)
pin=[0.]*9                           # initial
                                     # parameter guess
output=optimize.leastsq(residuals,pin)
```

```
pout=output[0]                           # optimized
                                         # parameters
def fun(x): return fitfun(x,pout)        # suitable for
                                         # plotting
```

Comments:
For simplicity a general least-squares optimization is used, although the optimization problem is linear here. The function to be fitted is $p_0 + p_1 \sin\phi + p_2 \cos\phi + p_3 \sin 2\phi + p_4 \cos 2\phi + p_5 \sin 3\phi + p_6 \cos 3\phi + p_7 \sin 4\phi + p_8 \cos 4\phi$. In view of the inaccuracy of the corrections, a fit with still higher harmonics is an overkill. The minimizer `leastsq` of the scipy package `optimize` is used.

Python code 7.1 (page 94) *Nonlinear least-squares fit, urease kinetics*

```
from scipy import optimize
S = array([30.,60.,100.,150.,250.,400.])
v = array([3.09,5.52,7.59,8.72,10.69,12.34])
```

A. Minimization using leastsq:

```
lsq = optimize.leastsq
def residuals(p):
    [vmax,Km]=p
    return v-vmax*S/(Km+S)
output = lsq(residuals,[15,105])
pout = output[0]
```

B. Minimization using fmin_powell:

```
def fun(S,p):
    [vmax,Km]=p
    return vmax*S/(Km+S)
def SSQ(p): return ((v-fun(S,p))**2).sum()
pin = [15,105]
pout = optimize.fmin_powell(SSQ,pin)
```

Comments:
The function `leastsq` requires as input an array of residues as a function of the parameters; it minimizes its sum of squares. The function `fmin_powell` adjusts the parameters in `fun` such that SSQ is a minimum. The new parameters `pout` are returned. The last line may be repeated with the new parameters as input. These minimization procedures do not need any derivatives. Check SSQ by the command `print SSQ(pout)`. In this example method A gives a more accurate result than method B.

Python code 7.2 (page 96) *Generate the cumulative probability for given* χ^2

```
from scipy import stats
cdf=stats.chi2.cdf
ppf=stats.chi2.ppf
```

Comments:
The function cdf (x,ν) gives the probability that χ^2, i.e., the sum of ν squares of random samples from a normal distribution, is less than x. For example, for 15 degrees of freedom, the probability of finding $\chi^2 \leq 10.5$ is given by
```
print cdf(10.5,15)
```
and the probability of finding $\chi^2 \geq 18.3$ is given by
```
1.-cdf(18.3,15)
```
The values of χ^2 for which the probabilities that the sum of 15 squares does not exceed χ^2 are 1,2,5,10 % are given by
```
ppf(array([1.,2.,5.,10.])*0.01,15)
```
The values of χ^2 for which the probabilities that the sum of 15 squares exceeds χ^2 are 1,2,5,10 % are given by
```
ppf(array([99.,98.,95.,90.])*0.01,15).
```

Python code 7.3 (page 103) *Generate and plot a contour for a two-dimensional function*

```
from scipy import *
from scipy import optimize
def contour(fxy,z,xycenter,xyscale=[1.,1.],\
  radius=0.05,nmax=500):
# construct contour f(x,y)=z by succession of
# circular intersects
# input:
#    fxy(x,y): defined function;
#    z:        level
#    xycenter: [xc,yc] point within contour
#    xyscale:  [xscale,yscale] approximate
#              coordinate ranges
#    radius:   radius of circle in units of
#              coordinate range
#    nmax:     maximum number of points (for open
#              contours)
    from scipy import optimize
    x0=xycenter[0]; y0=xycenter[1]
    xscale=xyscale[0]; yscale=xyscale[1]
```

```
            def funx(x):
                return fxy(x,y0)-z
            def funphi(phi):
                # uses xa,xb; dxs,dys (scaled)
                sinphi=sin(phi); cosphi=cos(phi)
                x=xa+(dxs*cosphi+dys*sinphi)*xscale
                y=ya+(-dxs*sinphi+dys*cosphi)*yscale
                return fxy(x,y)-z
            # find first point on x-axis
            xx=optimize.brentq(funx,x0,x0+5.*xscale)
            xlist=[xx]; ylist=[y0]
            # find second point
            dxs=radius; dys=0.
            xa=xx; ya=y0
            phi=optimize.brentq(funphi,-pi,0.)
            sinphi=sin(phi); cosphi=cos(phi)
            xb=xa+(dxs*cosphi+dys*sinphi)*xscale
            yb=ya+(-dxs*sinphi+dys*cosphi)*yscale
            xlist += [xb]; ylist += [yb]
            # find next point
            radsq=radius*radius
            dsq=4.*radsq
            n=0
            while (dsq>radsq) and (n<nmax):
                n +=1
                dxs=(xb-xa)/xscale
                dys=(yb-ya)/yscale
                xa=xb; ya=yb
                phi=optimize.brentq(funphi,-0.5*pi,0.5*pi)
                sinphi=sin(phi); cosphi=cos(phi)
                xb=xa+(dxs*cosphi+dys*sinphi)*xscale
                yb=ya+(-dxs*sinphi+dys*cosphi)*yscale
                xlist += [xb]; ylist += [yb]} \\
                dsq=((xb-xx)/xscale)**2+((yb-y0)/yscale)**2
            xlist += [xx]; ylist += [y0]
            data=array([xlist,ylist])
            return data
```

Comments:
This function produces an array of coordinate values $[x, y]$ along a contour
for which $f(x, y) = z$. Here $f(x, y)$ is a predefined function and z is a prescribed level. This array can be simply plotted by connecting the points with
straight lines, e.g.:
```
autoplotp(data,symbol=' ',lines=Black)
```

The points are generated as follows. The first point is located on a line parallel to the x-axis, starting from the point [xc,yc], searching in positive direction. Thus the input point [xc,yc] should be located inside the contour. The second point is searched on a half-circular (positive y) contour around the first point, with radius radius. Subsequent points are searched on a half-circular contour around the previous point, searching in the forward direction. Thus radius is the distance between subsequent points, which determines the resolution of the plot. The search is done in *scaled* x, y coordinates in order to prevent uneven distribution of points. The input xyscale is used for scaling: x-values are divided by xyscale[0] and y-values by xyscale[1]. You can use the total width of the plotted scales for xyscale, but the choice is not critical. The default radius 0.05 then means that the distance between points is 5 percent of the plot size. The optional parameter nmax limits the number of points generated on the contour, preventing infinite search along open contours. If a closed contour appears to be incomplete, either increase nmax or increase radius.

Python code 7.4 (page 104) *Generate a* $\Delta \chi^2 = 1$ *contour and derive uncertainties for the urease kinetics example*

Start from python code 7.1, which defines SSQ(p), p[0]= v_{max}; p[1]= K_m, pout = *best parameter values.*

```
from scipy import *
from plotsvg import *
S0=SSQ(pout)
def fxy(x,y): return 4./S0*(SSQ([x,y])-S0)
data=contour(fxy,1.,pout,xyscale=[0.4,7.])
# plot the contour:
f=Figure()
f.frame([[15.25,16.25],[105,125]])
f.plotp(data,lines=Black,symbol='')
f.show()
# compute sig1,sig2, rho from contour:
sig1=0.5*(data[0].max()-data[0].min())
sig2=0.5*(data[1].max()-data[1].min())
ratio=(data[0,0]-pout[0])/sig1
rho=sqrt(1.-ratio**2)
print 'sigma1= %5.2f, sigma2=%5.2f, rho=%5.2f' %\
      (sig1,sig2,rho)
```

Comments:
The function fxy defines $\Delta \chi^2$ as a function of the parameters. The input xyscale in the function contour (see python code 7.3) is taken as

estimates of the standard deviations. The contour data array `data` contains 122 points; you may increase the resolution by setting a smaller `radius`. The standard uncertainties are derived from the extrema of the contour data; the correlation coefficient is found from the x-intercept `data[0,0]` by using the rule that the intercept occurs at a fraction $\sqrt{1 - \rho^2}$ of the standard deviation.

Python code 7.5 (page 105) *Generate the covariance matrix by minimization (urease kinetics example)*

Start from python code 7.1, which defines `residuals(p)`, `SSQ(p)`, `p[0]=` v_{max}; `p[1]=` K_m. *We redo the minimization with full output:*

```
from scipy import optimize
lsq=optimize.leastsq
output = lsq(residuals,[15,105],full_output=1)
pout = output[0]
S0=SSQ(pout)
C=S0/(n-m)*output[1]
sig1=sqrt(C[0,0])
sig2=sqrt(C[1,1])
rho=C[0,1]/sig1/sig2
print 'sigma1= %5.2f, sigma2=%5.2f, rho=\%5.2f' %\
      (sig1,sig2,rho)
```

Comments:
The routine `leastsq` has a *full output* option, which produces as second element the covariance matrix C, be it without proper scaling. The output matrix is only equal to the covariance matrix if all standard uncertainties σ_y are equal to 1. Correct results are obtained if the output matrix is scaled by $S_0/(n - m)$.

Python code 7.6 (page 105) *Generate the covariance matrix from the B-matrix (urease kinetics example)*

First construct matrix B in general, given function `delchisq(p)`

```
from scipy import *
def matrixB(delchisq, delta):
# delchisq(delp) = chisq(p-p0)-chisq(p0)
# delta = array of test deviations
    m=len(delta)
    B=zeros((m,m))
```

```
d=zeros(m)
fun=zeros(m)
if (abs(delchisq(d)) > 1.e-8):
    print 'definition delchisq incorrect'
for i in range(m):
    di=delta[i]
    d[i]=di
    fun[i]=delchisq(d)
    B[i,i]=fun[i]/(di*di)
    for j in range(i):
        dj=delta[j]
        d[j]=dj
        funij=delchisq(d)
        B[j,i]=B[i,j]=0.5*(funij-fun[i]-\
        fun[j])/(di*dj)
        d[j]=0.
    d[i]=0.
return B
```

Start from python code 7.1, which defines residuals(p), SSQ(p), $p[0] = v_{max}$; $p[1] = K_m$, pout *= best parameter values. First construct B, then invert B and print results.*

```
delta=array([0.2,3.5]) # displacements near
                       # delchisq = 1
def delchisq(delp): return 4.*(SSQ(pout+delp)/\
    S0-1.)
B=matrixB(delchisq,delta)
from numpy import linalg
C=linalg.inv(B)
sig1=sqrt(C[0,0])
sig2=sqrt(C[1,1])
rho=C[0,1]/sig1/sig2
print 'sigma1=%5.2f, sigma2=%5.2f, rho=%5.2f' %\
    (sig1,sig2,rho)
```

Comments:
The construction of B proceeds by stepping delta[i] in all directions (which yields the diagonal elements) and stepping all pairs delta[i],delta[j] (which yields the off-diagonal elements). This is a simple procedure that could be made more sophisticated by involving steps in the opposite directions as well. Matrix inversion is done by the routine inv contained in the numpy module linalg.

Python code 7.7 (page 106) **Fit**: *a program that reports a general least-squares fit of a predefined function to a set of independent data.*

```
from scipy import *
from plotsvg import *
def fit(function,data,parin,figures=True):
    ,,,
Function. Non-linear least-squares fit of function
----------------------------------------------------
(x,par) to data
----------------
arguments:
  function          predefined function(x,par),
                    where x=independent variable
                    (called with an array x=data
                    [0]); par is a list of
                    parameters, e.g. [a,b]
  data              list (or 2D array) [x,y] or
                    [x,y,sig]. sig contains
                    standard deviations of y
                    (if known). If sig is given, a
                    chi-squared test is done; if
                    not given, equal weights are
                    assumed.
  parin             list of initial values for the
                    parameters, e.g. [0.,1.]
  figures=True      if True, two figures are
                    produced and displayed
returns:            [parout,sigma] (final parameters
                    with s.d.)

Remarks:
  The sum of weighted square deviations
  chisq=sum(((y-function(x))/sig)**2) [or, if no
  sig is given, SSQ=sum(((y-function(x)))**2)] is
  minimized by the nonlinear Scipy routine
  leastsq., using function values only. After
  successful determination of the best fit,
  uncertainties (s.d. and correlation
  coefficients) are computed, including the full
  covariance matrix. Plots of the fit and the
  residuals are produced.
```

```
Example: fit exponential function to data [x,y]
with sd sig:
  >>>def f(x,par):
         [a,k,c]=par
         return a*exp(-k*x)+c
  >>>[[a,k,b],[siga,sigk,sigc]]=fit(f,[x,y,sig],
  [1.,0.1,0.])
'''
   import os
   from scipy import optimize,stats
   lsq=optimize.leastsq
   if (len(data)==2):
       weights=False
   elif (len(data)==3):
       weights=True
   else:
       print 'ERROR: data should contain 2 or\
              3 items'
       print 'fit aborted'
       return 0
   m=len(parin)
   x=array(data[0])
   n=len(x)
   y=array(data[1])
   if (len(y)!=n):
       print 'ERROR: x and y have unequal length'
       print 'fit aborted'
       return 0
   if weights:
       sig=array(data[2])
       if (len(sig)!=n):
           print 'ERROR: x and sig have unequal\
                  length'
           print 'fit aborted'
           return 0
       def residuals(p): return (y-\
         function(x,p))/sig
   else:
       def residuals(p): return (y-function(x,p))
   def SSQ(p): return (residuals(p)**2).sum()
   SSQ0=SSQ(parin)
   # print results after minimization:
   print '\n Report on least-squares parameter\
          fit'
```

```
if weights:
    print 'chisq = sum of square reduced dev.\
        (y-f(x))/sig'
else:
    print 'SSQ = sum of square deviations\
        (y-f(x))'
print '\nnr of data points:        %5d' % (n)
print    'nr of parameters:        %5d' % (m)
print    'nr of degrees of freedom: %5d' % (n-m)
print '\nInitial values of parameters: '
print parin
if weights:
    print 'Initial chisq = %10.6g' % (SSQ0)
else:
    print 'Initial SSQ = %10.6g' % (SSQ0)
output=lsq(residuals,parin,full_output=1)
parout=output[0]
SSQout=SSQ(parout)
print 'Results after minimization:'
if weights:
    print 'Final chisq = %10.6g' % (SSQout)
else:
    print 'Final SSQ = %10.6g' % (SSQout)
print 'Final values of parameters'
print parout
# covariance matrix C
C=SSQout/(n-m)*output[1]
sigma=arange(m,dtype=float)
for i in range(m): sigma[i]=sqrt(C[i,i])
print 'Standard inaccuracies of parameters,:'
print sigma
print '\nMatrix of covariances'
print C
SR=zeros((m,m),dtype=float)
for i in range(m):
    SR[i,i]=sigma[i]
    for j in range(i+1,m):
        SR[j,i]=SR[i,j]=C[i,j]/(sigma[i]*
        sigma[j])
print '\nMatrix of sd (diagonal) and corr.
        coeff. (off-diag)'
print SR
# chisq analysis
if weights:
```

```
nu=n-m
F=stats.chi2.cdf(SSQout,nu)
print '\nChi-square analysis:'
print 'chi^2 (sum of weighted square\
    deviations) =%10.6g' % (SSQout)
print 'cum. prob. for chi^2  = %5.3g %%' %\
    (100.*F)
if (F<.1):
    print "chi^2 is low! Did you\
        overestimate the supplied\
        sigma's?"
    print 'Or did you fit the original\
        data too closely with too many\
        parameters?'
elif (F>.9):
    print "chi^2 is high! Did you neglect\
        an error source in the supplied\
        sigma's?"
    print 'Or did the data result from a\
        bad fitting procedure?'
else:
    print 'cum. probab. of chi^2 is\
        reasonable (between 10% and 90%).'
    print "The spread in the data agrees\
        with the supplied sigma's"
# produce two plots (data and fitting curve;
#   residuals)
if figures:
    xmin=x.min(); xmax=x.max()
    ymin=y.min(); ymax=y.max()
    if weights:
        maxsigy=sig.max()
        ymin=ymin-maxsigy
        ymax=ymax+maxsigy
    y1=1.05*ymin-0.05*ymax; y2=1.05*ymax
    -0.05*ymin
    f=Figure()
    f.frame([[xmin,xmax],[y1,y2]],title=\
        'Least-squares fit')
    if weights:
        f.plotp([x,y],symbolfill=Blue,ybars=\
        sig, barcolor=Blue)
    else:
        f.plotp([x,y],symbolfill=Blue)
```

```
def fun(x): return function(x,parout)
f.plotf(fun,color=Red)
f.show(filename='figfit.svg')
os.startfile('figfit.svg')
print 'figfit.svg is now displayed by your\
        browser'
residuals=y-fun(x)
minres=residuals.min(); maxres=residuals.
max()
if weights:
    minres=minres-maxsigy
    maxres=maxres+maxsigy
y1=1.05*minres-0.05*maxres; y2=1.05*maxres\
    -0.05*minres
f=Figure()
f.size=[5500,3400]
f.frame([[xmin,xmax],[y1,y2]],
title="residuals")
if weights:
    f.plotp([x,residuals],symbolfill=Blue,\
    ybars=sig, barcolor=Blue)
else:
    f.plotp([x,residuals],symbolfill=Blue)
f.show(filename='figresiduals.svg')
os.startfile('figresiduals.svg')
print 'figresiduals.svg is now displayed\
        by your browser'
return [parout,sigma]
```

Comments:
This program can be downloaded from www.hjcb.nl/python. Look for recent updates. Two plots are automatically generated and displayed by the standard browser. Make sure that the .svg mime type starts your SVG-enabled browser.

Python code 7.8 (page 108) *Compute various sum of squared deviations and perform an F-test on the urease kinetics example*

Start from python code 7.1 and code 7.5, which defines independent variable S, *dependent variable* v *and best parameters* pout.

```
y=S
def fun(x,p): return p[0]*x/(p[1]+x)
def ssq(x): return (x**2).sum()
```

```
f=fun(v,pout)
SST=ssq(y-y.mean())
SSR=ssq(f-f.mean())
SSE=ssq(y-f)
Fratio=SSR/(SSE/4.)
from scipy import stats
Fcum=stats.f.cdf(Fratio,1,4)
print 'SST=%7.3f SSR=%7.3f SSE=%7.3f' % (SST,SSR,\
     SSE)
print 'Fratio=%7.3f Fcum=%7.3f' % (Fratio,Fcum)
```

Comments:

The function $ssq(x)$ computes the sum of squares of the elements of a
1D-array x. The array f gives the best-fitted function values. The function
$f.cdf(Fratio, nu1, nu2)$ of the scipy module stats gives the
cumulative F-distribution.

Python code A5.1 (page 150) *Generate pdf of sum of n homogeneously
distributed random numbers*

```
from scipy import fftpack
def symmetrize(x): # adds mirror to x
    n=alen(x)
    half=n/2
    for i in range(1,half): x[n-i] += x[i]
    return 1
def FT(Fx,delx):    # produces real FT of symmetric
                    # Fx
    Gy=fftpack.fft(Fx).real*delx
    return Gy
def IFT(Gy,delx):   # produces real inverse FT of
                    # symmetric Gy
    Fx=fftpack.ifft(Gy).real/delx
    return Fx
n=10               # number of random numbers to be
                   # added
a=sqrt(3./n)       # [-a,a] is range of random
                   # numbers
nft=4096           # array length for FT
xm=50.             # maximum of x-scale
delx=2.*xm/nft     # delta x between points in Fx
dely=pi/xm         # delta y between points in Gy
ym=nft*dely/2.     # maximum of y-scale
Fx=zeros((nft),dtype=float)
```

```
# set rectangular function Fx:
for i in range(int(a/delx)): Fx[i]=0.5/a
symmetrize(Fx)      # this makes the FT real
corr=1./Fx.sum()/delx  # correction to make
                       # integral exactly = 1
Fx=Fx*corr
Gy=FT(Fx,delx)      # Fourier transform of
                    # rectangular function
Gyn=Gy**n           # FT of convolution of 10
                    # rectangular functions
Fxn=IFT(Gyn,delx)   # Inverse FT gives the
                    # convolution function
m=4./delx           # [-4,4] is interesting plot
                    # range
yn=concatenate((Fxn[-m:],Fxn[:m]))
                    # yn is the useful output
```

Comments:
This example computes the probability density function of the sum of $n = 10$ homogeneously distributed random numbers from an interval $[-a, a\rangle$, where a is chosen such that the resulting variance of the sum equals 1. Such a distribution is a convolution of 10 rectangular functions, which is most easily computed by inverse FT of the n-th power of the FT of the original rectangular function. In the last lines the result is recast into a smaller range, symmetrical around zero $(-4 < x < 4)$.

> **Python code A7.1** (page 157) *Variance of the mean by block averages*

```
def block(data,n):
# block-average data in blocks of length n
# data: input [x,y] (x,y: 1D-arrays of same length)
# n: number of points in each block
# returns array of block averages of both x and y
    ntot=len(data[0])
    nnew=ntot/n
    x=zeros(nnew,dtype=float)
    y=zeros(nnew,dtype=float)
    for i in range(nnew):
        x[i]=sum(data[0][i*n:(i+1)*n])/float(n)
        y[i]=sum(data[1][i*n:(i+1)*n])/float(n)
    return [x,y][1ex]
def blockerror(data,blocksize=[10,20,40,60,80,100,\
    125]):
```

```
# make list of s.d of the mean for given blocksizes
# data: input [x,y] (x,y: 1D-arrays of same length)
# blocksize: list of lengths of blocks,
#    assuming independent block averages
# returns [blocksize, stderror, ybars]
# ybars is rms inaccuracy of stderror
    n=len(data[1])
    delt=(data[0][-1]-data[0][0])/float(n-1)
    xout=[]
    yout=[]
    ybars=[]
    for nb in blocksize:
        xyblock=block(data,nb)
        number = len(xyblock[1])
        std=xyblock[1].std()
        stderror = std/sqrt(number-1.)
        xout += [nb*delt]
        yout += [stderror]
        ybars += [stderror/sqrt(2.*(number-1.))]
    return [xout,yout,ybars]
```

Comments:
It is assumed that a set of data=[x,y] is available (*x* and *y* being arrays).
The function block(data,n) returns a new set of data consisting of the
averages over blocks of length *n*. The blocks start at the first data item; if the
number of data points does not fit an integer number of blocks, the remaining
points are not used. The function blockerror calls the function block
for each of the elements in the optional argument blocksize. For each
blocksize it computes the standard error in the mean and outputs it as yout.
The output values xout are the block sizes expressed in units of *x*. The
output values ybars are the standard deviations expected for yout on the
basis of the limited number of averages; it can be used to draw error bars in
a plot of the output data.

PART IV

Scientific data

Chi-squared distribution

Probability distribution sum of squares

$x_1, x_2, \ldots, x_\nu$ are independent, normally distributed variables with $E\{x_i\} = 0$ and $E\{x_i^2\} = 1$; $\nu =$ number of *degrees of freedom*; $\chi^2 = \sum_{i=1}^{\nu} x_i^2$. The probability density function of χ^2 is:

$$f(\chi^2|\nu)\,d\chi^2 = [2^{\nu/2}\,\Gamma(\tfrac{\nu}{2})]^{-1}(\chi^2)^{\nu/2-1}\,\exp[-\chi^2/2]\,d\chi^2.$$

Moments of $f(\chi^2|\nu)$:

mean	$\mu = E\{\chi^2\}$	$= \nu$
variance	$\sigma^2 = E\{(\chi^2 - \mu)^2\}$	$= 2\nu$
skewness	$\gamma_1 = E\{(\chi^2 - \mu)^3/\sigma^3\}$	$= 2\sqrt{(2/\nu)}$
excess	$\gamma_2 = E\{(\chi^2 - \mu)^4/\sigma^4 - 3\}$	$= 12/\nu$

Special cases

ν	$f(\chi^2|\nu)$
1	$(2\pi)^{-1/2}\chi^{-1}\exp[-\chi^2/2]$
2	$\tfrac{1}{2}\exp[-\chi^2/2]$
3	$(2\pi)^{-1/2}\chi\exp[-\chi^2/2]$
∞	$(4\pi\nu)^{-1/2}\exp[-(\chi^2 - \nu)^2/(4\nu)]$
	normal with var $= 2\nu$

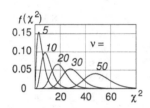

Relation to Poisson distribution (ν even)

$$1 - F(\chi^2|\nu) =$$
$$= \sum_{j=0}^{c-1} e^{-m}m^j/j!,$$
$$c = \tfrac{1}{2}\nu \; m = \tfrac{1}{2}\chi^2).$$

Cumulative χ^2-distribution

$F(\chi^2|\nu) =$ probability that sum of squares $< \chi^2$:

$$F(\chi^2|\nu) = \int_0^{\chi^2} f(S|\nu)\,dS. \quad \textit{See table p. 2.}$$

Probability that χ^2 is exceeded is $1 - F(\chi^2)$.

DATA SHEET

Chi-squared distribution

Values of χ^2 for 1%, 10%, 50%, 90%, and 99%

$F = \nu$	0.01	0.10	0.50	0.90	0.99
1	0.000	0.016	0.455	2.706	6.635
2	0.020	0.211	1.386	4.605	9.210
3	0.115	0.584	2.366	6.251	11.35
4	0.297	1.064	3.357	7.779	13.28
5	0.554	1.610	4.351	9.236	15.09
6	0.872	2.204	5.348	10.65	16.81
7	1.239	2.833	6.346	12.02	18.48
8	1.646	3.490	7.344	13.36	20.09
9	2.088	4.168	8.343	14.68	21.67
10	2.558	4.865	9.342	15.99	23.21
11	3.053	5.578	10.34	17.28	24.73
12	3.571	6.304	11.34	18.55	26.22
13	4.107	7.042	12.34	19.81	27.69
14	4.660	7.790	13.34	21.06	29.14
15	5.229	8.547	14.34	22.31	30.58
20	8.260	12.44	19.34	28.41	37.57
25	11.52	16.47	24.34	34.38	44.31
30	14.95	20.60	29.34	40.26	50.89
40	22.16	29.05	39.34	51.81	63.69
50	29.71	37.69	49.34	63.17	76.15
60	37.49	46.46	59.34	74.40	88.38
70	45.44	55.33	69.33	85.53	100.4
80	53.54	64.28	79.33	96.58	112.3
90	61.75	73.29	89.33	107.6	124.1
100	70.07	82.36	99.33	118.5	135.8
∞	$\nu - a$	$\nu - b$	ν	$\nu + b$	$\nu + a$
	$a = 3.290\sqrt{\nu}$			$b = 1.812\sqrt{\nu}$	

F-distribution

F-distribution

Meaning of variable: F-ratio = ratio of mean squared deviations of two groups of samples.

$$F_{\nu_1,\nu_2} = \frac{\mathrm{MSD}_1}{\mathrm{MSD}_2} = \frac{\sum(\Delta y_{1i})^2/\nu_1}{\sum(\Delta y_{2i})^2/\nu_2}.$$

F-test: yields (cumulative) probability that both groups come from distributions with the same variance.

Probability density function:

$$f(F_{\nu_1,\nu_2}) = \frac{\Gamma\left(\frac{\nu_1+\nu_2}{2}\right)}{\Gamma\left(\frac{\nu_1}{2}\right)\Gamma\left(\frac{\nu_2}{2}\right)} \nu_1^{\nu_1/2}\nu_2^{\nu_2/2} F^{(\nu_1-2)/2}(\nu_2+\nu_1 F)^{-(\nu_1+\nu_2)/2}.$$

Cumulative distribution function:

$F(F_{\nu_1,\nu_2}) = \int_{-\infty}^{F} f(F')\,dF'$

$1 - F(F_{\nu_1,\nu_2}) = \int_{F}^{\infty} f(F')\,dF'$

mean: $m = \nu_2/(\nu_2 - 2)$, $\nu_2 > 2$

variance: $\sigma^2 = 2\nu_2^2(\nu_1 + \nu_2 - 2)/[\nu_1(\nu_2 - 2)^2(\nu_2 - 4)]$, $\nu_2 > 4$.

Reflexive relation:

$F(F_{\nu_1,\nu_2}) = 1 - F(1/F_{\nu_2,\nu_1})$

e.g. $F_{10,5} = 4.74$ at the 95% level; then $F_{5,10} = 1/4.74 = 0.21$ at the 5% level.

Therefore tables can be restricted to F-ratios > 1.

Use in ANOVA (analysis of variance) in regression

Given: n data (x_i, y_i), $i = 1, \ldots, n$. Fit $f_i = ax_i + b$ by linear regression. The total sum of squared deviations SST can be divided into SSR (regression SSQ, explained by the model) and SSE (remaining error). $\nu = $ nr of degrees of freedom:

SST $(\nu = n - 1)$ = SSR $(\nu = 1)$ + SSE $(\nu = n - 2)$

SST $= \sum(y_i - \langle y \rangle)^2$; SSR $= \sum(f_i - \langle y \rangle)^2$; SSE $= \sum(y_i - f_i)^2$

Perform F-test on $F_{1,n-2} = [\mathrm{SSR}/1]/[\mathrm{SSE}/(n-2)]$.

Remark: For regression with m parameters:

Perform F-test on $F_{m-1,n-m} = [\mathrm{SSR}/(m-1)]/[\mathrm{SSE}/(n-m)]$.

F-distribution

F-distribution, percentage points 95% and 99%

$F(F_{v_1,v_2}) = \mathbf{0.95}$

If $(\sum y_{1i}^2/v_1)/(\sum y_{2i}^2/v_2)$ exceeds the F-ratio F_{v_1,v_2} given in the table, the probability is $< 5\%$ that y and z are samples from distributions with equal variance.

v_1 v_2	1	2	3	4	5	7	10	20	50	∞
2	18.5	19.0	19.2	19.3	19.3	19.4	19.4	19.5	19.5	19.5
3	10.1	9.55	9.28	9.12	9.01	8.89	8.79	8.66	8.58	8.53
4	7.71	6.94	6.59	6.39	6.26	6.09	5.96	5.80	5.70	5.63
5	6.61	5.79	5.41	5.19	5.05	4.88	4.74	4.56	4.44	4.36
7	5.59	4.74	4.35	4.12	3.97	3.79	3.64	3.44	3.32	3.23
10	4.96	4.10	3.71	3.48	3.33	3.14	2.98	2.77	2.64	2.54
20	4.35	3.49	3.10	2.87	2.71	2.51	2.35	2.12	1.97	1.84
50	4.03	3.18	2.79	2.56	2.40	2.20	2.03	1.78	1.60	1.44
∞	3.84	3.00	2.61	2.37	2.21	2.01	1.83	1.57	1.35	1.00

$F(F_{v_1,v_2}) = \mathbf{0.99}$

v_1 v_2	1	2	3	4	5	7	10	20	50	∞
2	98.5	99.0	99.2	99.3	99.3	99.4	99.4	99.5	99.5	99.5
3	34.1	30.8	29.5	28.7	28.2	27.7	27.2	26.7	26.4	26.1
4	21,2	18.0	16.7	16.0	15.5	15.0	14.6	14.0	13.7	13.5
5	16.3	13.3	12.1	11.4	11.0	10.5	10.1	9.55	9.24	9.02
7	12.3	9.55	8.45	7.85	7.46	6.99	6.62	6.16	5.86	5.65
10	10.0	7.56	6.55	5.99	5.64	5.20	4.85	4.41	4.12	3.91
20	8.10	5.85	4.94	4.43	4.10	3.70	3.37	2.94	2.64	2.42
50	7.17	5.06	4.20	3.72	3.41	3.02	2.70	2.27	1.95	1.68
∞	6.63	4.61	3.78	3.32	3.02	2.64	2.32	1.88	1.53	1.00

Least-squares fitting

General least-squares fitting

Sum of weighted squared deviations

a. *Uncorrelated data*

Given n measured values y_i, $i = 1, \ldots n$, we seek m parameters $\hat{\theta}_k$, $k = 1 \ldots m$, $m < n$; such that:

$$S = \sum_{i=1}^{n} w_i(y_i - f_i)^2 \text{ minimal}$$

$f_i(\theta_1, \ldots \theta_m)$ are functions of parameters. For the minimum: $S(\hat{\theta}) = S_0$. Both y_i and f_i can be functions of one or more independent variables.

The *residuals* $\varepsilon_i = y_i - f_i$ are supposed to be samples from a random distribution with properties: $E[\varepsilon_i] = 0$; $E[\varepsilon_i \varepsilon_j] = \sigma_i^2 \delta_{ij}$.

The *weight factors* w_i should be proportional to σ_i^{-2}.

If the variances σ_i of the deviations are known, a chi-squared test can be carried out on $\chi_0^2 = \min \sum_{i=1}^{n} [(y_i - f_i)/\sigma_i]^2$, for $\nu = n - m$ degrees of freedom.

b. *Correlated data*

$S = \sum_{i,j=1}^{n} w_{ij}(y_i - f_i)(y_j - f_j)$ minimal, with $\varepsilon_i = y_i - f_i$ samples from a random distribution with properties: $E[\varepsilon_i] = 0$; $E[\varepsilon_i \varepsilon_j] = \Sigma_{ij}$. $\boldsymbol{\Sigma}$ is the covariance matrix of the measured values. The matrix $\boldsymbol{W}$ of weight factors should be proportional to $\boldsymbol{\Sigma}^{-1}$.

Parameter covariances Likelihood of $\boldsymbol{\theta}$ is proportional to $\exp\left[-\frac{1}{2}\chi^2(\boldsymbol{\theta})\right]$. Since $E[\chi_0^2] = n - m$, $\chi^2(\boldsymbol{\theta})$ is estimated by scaling S:

$\hat{\chi}^2(\boldsymbol{\theta}) = (n - m)S(\boldsymbol{\theta})/S_0 = n - m + (\boldsymbol{\Delta\theta})^T \boldsymbol{B} \boldsymbol{\Delta\theta}$, where $\boldsymbol{\Delta\theta} = \boldsymbol{\theta} - \hat{\boldsymbol{\theta}}$. The expectation of the parameter covariance matrix $\boldsymbol{C} = E[(\boldsymbol{\Delta\theta})(\boldsymbol{\Delta\theta})^T]$ is given by:

$$\boldsymbol{C} = \boldsymbol{B}^{-1}.$$

$\sigma_k = \sqrt{C_{kk}}$; $\rho_{kl} = C_{kl}/(\sigma_k \sigma_l)$.

DATA SHEET

Least-squares fitting

Linear in the parameters

When f_i are linear functions of $\boldsymbol{\theta}$:
$f_i(\theta) = \sum_k A_{ik}\theta_k$; $\boldsymbol{f} = \boldsymbol{A}\,\boldsymbol{\theta}$ (general: $A_{ik} = \partial f_i/\partial\theta_k$)

$$S = (\boldsymbol{y} - \boldsymbol{f})^{\mathrm{T}}\boldsymbol{W}(\boldsymbol{y} - \boldsymbol{f})\ \text{minimal}$$

$$\text{for }\hat{\boldsymbol{\theta}} = (\boldsymbol{A}^{\mathrm{T}}\boldsymbol{W}\boldsymbol{A})^{-1}\boldsymbol{A}^{\mathrm{T}}\boldsymbol{W}\boldsymbol{y},$$

where $W_{ij} \propto \sigma_i^{-2}\delta_{ij}$ (uncorrelated data).
$S(\hat{\boldsymbol{\theta}}) = S_0$
Expectation of parameter covariance matrix $\boldsymbol{C} = E[(\boldsymbol{\Delta\theta})(\boldsymbol{\Delta\theta})^{\mathrm{T}}]$ is given by:

$$\boldsymbol{C} = [S_0/(n - m)](\boldsymbol{A}^{\mathrm{T}}\boldsymbol{W}\boldsymbol{A})^{-1}.$$

Special case: linear function:
$f_i = f(x_i) = ax_i + b$ (a and b parameters):

$$a = \langle(\Delta x)(\Delta y)\rangle/\langle(\Delta x)^2\rangle;\ b = \langle y\rangle - a\langle x\rangle.$$

Here $\langle\ \rangle$ are *weighted* averages, such as:

$\langle\xi\rangle = (1/w)\sum_{i=1}^{n} w_i\,\xi_i;\ w = \sum_{i=1}^{n} w_i.$
$\Delta x = x - \langle x\rangle;\ \Delta y = y - \langle y\rangle.$

Expectation of (co)variances of a and b:
$E[(\Delta a)^2] = \sigma_a^2 = S_0/[n(n-2)\langle(\Delta x)^2\rangle]$
$E[(\Delta b)^2] = \sigma_b^2 = \langle x^2\rangle\sigma_a^2$
$E[\Delta a\Delta b] = -\langle x\rangle\sigma_a^2;\ \rho_{ab} = -\langle x\rangle\sigma_a/\sigma_b$
N.B.: a and b are uncorrelated if $\langle x\rangle = 0$.

Correlation coefficient r of x and y:

$$r = \frac{\langle(\Delta x)(\Delta y)\rangle}{\sqrt{\langle(\Delta x)^2\rangle}\sqrt{\langle(\Delta y)^2\rangle}} = a\left(\frac{\langle(\Delta x)^2\rangle}{\langle(\Delta y)^2\rangle}\right)^{1/2}.$$

Normal distribution

One-dimensional Gauss function

Probability density function:

$$f(x)\,dx = (\sigma\sqrt{2\pi})^{-1}\exp[-(x-\mu)^2/(2\sigma^2)]\,dx$$

$\mu = mean,$
$\sigma^2 = variance,$
$\sigma = standard\ deviation.$
Standard form:
$f(z) = (1/\sqrt{2\pi})\exp(-z^2/2),$
$z = (x-\mu)/\sigma.$

Characteristic function: $\Phi(t) = \exp\left(-\frac{1}{2}\sigma^2 t^2\right)\exp(i\mu t).$

Central moments $\mu_n = \int_{-\infty}^{\infty}(x-\mu)^n f(x)\,dx,$

$\mu_m = 0$ for m even, $\mu_{2n} = \sigma^{2n}\times 1\times 3\times 5\times(2n-1),$
$\mu_2 = \sigma^2,\ \mu_4 = 3\sigma^4,\ \mu_6 = 15\sigma^6,\ \mu_8 = 105\sigma^8.$
$skewness = 0,\ excess = 0.$

Cumulative distribution function:

$$F(x) = \int_{-\infty}^{x} f(x')\,dx' = \frac{1}{2}\{1 + \mathrm{erf}\,(x/\sigma\sqrt{2})\},$$
$$1 - F(x) = F(-x) = \int_{x}^{\infty} f(x')\,dx' = \frac{1}{2}\,\mathrm{erfc}\,(x/\sigma\sqrt{2}).$$

z	$f(z)$	$F(-z)$	z	$f(z)$	$F(-z)$
0.0	0.3989	0.5000	1.4	1.497e-01	8.076e-02
0.1	0.3970	0.4602	1.6	1.109e-01	5.480e-02
0.2	0.3910	0.4207	1.8	7.895e-02	3.593e-02
0.3	0.3814	0.3821	2.0	5.399e-02	2.275e-02
0.4	0.3683	0.3446	2.5	1.753e-02	6.210e-03
0.5	0.3521	0.3085	3.0	4.432e-03	1.350e-03
0.6	0.3332	0.2743	3.5	8.727e-04	2.326e-04
0.7	0.3123	0.2420	4.0	1.338e-04	3.167e-05
0.8	0.2897	0.2119	5.0	1.487e-06	2.866e-07
0.9	0.2661	0.1841	7.0	9.135e-12	1.280e-12
1.0	0.2420	0.1587	10	7.695e-23	7.620e-24
1.2	0.1942	0.1151	15	5.531e-50	3.671e-51

large z : $F(-z) = 1 - F(z) \approx \frac{f(z)}{z}\left(1 - \frac{1}{z^2+2} + \cdots\right)$

Normal distribution

Multivariate Gauss functions

General n-dimensional form:
$$f(\mathbf{x})\, d\mathbf{x} = (2\pi)^{-n/2} (\det \mathbf{W})^{1/2} \exp\left[-\tfrac{1}{2}(\mathbf{x} - \mu)^{\mathrm{T}} \mathbf{W}(\mathbf{x} - \mu)\right] d\mathbf{x},$$
where $\mathbf{W}$ is the *weight matrix*. $\mathbf{W} = \mathbf{C}^{-1}$.
$\mathbf{C} \overset{\text{def}}{=} E[(\mathbf{x} - \mu)(\mathbf{x} - \mu)^{\mathrm{T}}]$ is the *covariance matrix*.

Bivariate normal distribution:
$$\mathbf{C} = \begin{pmatrix} \sigma_x^2 & \rho\sigma_x\sigma_y \\ \rho\sigma_x\sigma_y & \sigma_y^2 \end{pmatrix}; \quad \rho \text{ is the } correlation\ coefficient.$$

$$\mathbf{W} = \frac{1}{1-\rho^2}\begin{pmatrix} \sigma_x^{-2} & -\rho/(\sigma_x\sigma_y) \\ -\rho/(\sigma_x\sigma_y) & \sigma_y^{-2} \end{pmatrix}$$

$$f(x,y)\, dx\, dy = \frac{1}{2\pi\,\sigma_x\sigma_y\sqrt{1-\rho^2}} \exp\left[-\frac{z^2}{2(1-\rho^2)}\right] dx\, dy;$$

$$z^2 = \frac{(x-\mu_x)^2}{\sigma_x^2} - 2\frac{\rho(x-\mu_x)(y-\mu_y)}{\sigma_x\sigma_y} + \frac{(y-\mu_y)^2}{\sigma_y^2}.$$

Standard form:
$\mu_x = \mu_y = 0;\ \sigma_x = \sigma_y = 1;$
$r^2 = x^2 - 2\rho x y + y^2$ is equation
for ellipse at $+45°$ for $\rho > 0$
or $-45°$ for $\rho < 0$. Half major
axis $a = r/\sqrt{1 - |\rho|}$; half minor
axis $b = r/\sqrt{1 + |\rho|}$. Cumulative
probability integrated over ellipse:
$1 - \exp\left[-\tfrac{1}{2}r^2/(1 - \rho^2)\right]$.

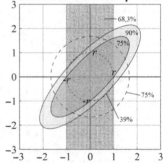

Bivariate Gaussian $\rho = 0.8$

Marginal distr.: $f_x(x) = (\sigma_x\sqrt{2\pi})^{-1} \exp\left[-\tfrac{1}{2}((x - \mu_x)/\sigma_x)^2\right]$,
Conditional distribution:

$$f(x|y) = \frac{1}{\sigma_x\sqrt{2\pi(1-\rho^2)}} \exp\left[-\frac{\{x - \mu_x - \rho(\sigma_x/\sigma_y)(y - \mu_y)\}^2}{2\sigma_x^2(1-\rho^2)}\right].$$

Conditional expectation: $E[x|y] = \mu_x + \rho(\sigma_x/\sigma_y)(y - \mu_y)$.

Normal distribution

Probability that ≥1 ∈ n samples exceeds an interval

Probability that at least one out of n (independent, normally distributed) samples falls outside the interval $(\mu - d, \mu + d)$:

$$\Pr\{\geq 1; n, d\} = 1 - [1 - 2F(-d/\sigma)]^n \text{ (double-sided)}$$

$n \downarrow$ $d/\sigma \rightarrow$	1.5	2	2.5	3	3.5	4
1	0.134	0.046	0.012	0.0027	4.7e-4	6.3e-5
2	0.249	0.089	0.025	0.0054	9.3e-4	1.3e-4
3	0.350	0.130	0.037	0.0081	0.0014	1.9e-4
4	0.437	0.170	0.049	0.0108	0.0019	2.5e-4
5	0.512	0.208	0.061	0.0134	0.0023	3.2e-4
6	0.577	0.244	0.072	0.0161	0.0028	3.8e-4
7	0.634	0.278	0.084	0.0187	0.0033	4.4e-4
8	0.683	0.311	0.095	0.0214	0.0037	5.1e-4
9	0.725	0.342	0.106	0.0240	0.0042	5.7e-4
10	0.762	0.372	0.117	0.0267	0.0046	6.3e-4
12	0.821	0.428	0.139	0.0319	0.0056	7.6e-4
15	0.884	0.503	0.171	0.0397	0.0070	9.5e-4
20	0.943	0.606	0.221	0.0526	0.0093	0.0013
25	0.972	0.688	0.268	0.0654	0.0116	0.0016
30	0.986	0.753	0.313	0.0779	0.0139	0.0019
40	0.997	0.845	0.393	0.102	0.0184	0.0025
50	0.999	0.903	0.465	0.126	0.0230	0.0032
70	1.000	0.962	0.583	0.172	0.0321	0.0044
100	1.000	0.991	0.713	0.237	0.0455	0.0063
150	1.000	0.999	0.847	0.333	0.0674	0.0095
200	1.000	1.000	0.918	0.418	0.0889	0.0126
300	1.000	1.000	0.976	0.556	0.130	0.0188
400	1.000	1.000	0.993	0.661	0.167	0.0250
500	1.000	1.000	0.998	0.741	0.208	0.0312

horizontal lines mark 5% level

Normal distribution

Probability that ≥1 ∈ *n* samples exceeds a value

Probability that at least one out of n (independent, normally distributed) samples is $> \mu + d$ *(or ...* $< \mu - d$*):*

$$\Pr\{\geq 1; n, d\} = 1 - [1 - F(-d/\sigma)]^n \text{ (single-sided)}$$

$n \downarrow$	$d/\sigma \rightarrow$ 1.5	2	2.5	3	3.5	4
1	0.067	0.023	0.0062	0.0014	2.3e-4	3.2e-5
2	0.129	0.045	0.012	0.0027	4.7e-4	6.3e-5
3	0.187	0.067	0.019	0.0040	6.9e-4	9.5e-5
4	0.242	0.088	0.025	0.0054	9.3e-4	1.3e-5
5	0.292	0.109	0.031	0.0067	0.0012	1.6e-4
6	0.340	0.129	0.037	0.0081	0.0014	1.9e-4
7	0.384	0.149	0.043	0.0094	0.0016	2.2e-4
8	0.425	0.168	0.049	0.011	0.0019	2.5e-4
9	0.463	0.187	0.055	0.012	0.0021	2.9e-4
10	0.499	0.206	0.060	0.013	0.0023	3.2e-4
12	0.564	0.241	0.072	0.016	0.0028	3.8e-4
15	0.646	0.292	0.089	0.020	0.0035	4.8e-4
20	0.749	0.369	0.117	0.027	0.0046	6.3e-4
25	0.823	0.438	0.144	0.033	0.0058	7.9e-4
30	0.874	0.499	0.170	0.038	0.0070	9.5e-4
40	0.937	0.602	0.221	0.053	0.0093	0.0013
50	0.968	0.684	0.268	0.065	0.012	0.0016
70	0.992	0.800	0.353	0.090	0.016	0.0022
100	0.999	0.900	0.464	0.126	0.023	0.0032
150	1.000	0.968	0.607	0.183	0.034	0.0047
200	1.000	0.990	0.712	0.237	0.045	0.0063
300	1.000	0.999	0.846	0.333	0.067	0.0095
400	1.000	1.000	0.917	0.417	0.089	0.0126
500	1.000	1.000	0.956	0.491	0.110	0.0157

horizontal lines mark 5% level

Physical constants

(between parentheses: standard deviation)

velocity of light	$c = 299\,792\,458$ m/s (exact)
magnetic constant	$\mu_0 = 4\pi \times 10^{-7}$ N/A^2 (exact)
	$= 1.256\,637\,0614\ldots \times 10^{-6}$
electric constant	$\varepsilon_0 = 1/\mu_0 c^2$ (exact)
	$= 8.854\,187\,817\ldots \times 10^{-12}$ F/m
characteristic impedance	$Z_0 = \sqrt{\mu_0/\epsilon_0} = \mu_0 c$ (exact)
vacuum	$= 376.730\,313\,461\ldots \Omega$
Planck constant	$h = 6.626\,068\,96(33) \times 10^{-34}$ J s
Dirac constant $h/2\pi$	$\hbar = 1.054\,571\,628(53) \times 10^{-34}$ J s
gravitational constant	$G = 6.674\,28(67) \times 10^{-11}$ m^3 kg^{-1} s^{-2}
elementary charge	$e = 1.602\,176\,487(40) \times 10^{-19}$ C
mass electron	$m_e = 9.109\,382\,15(45) \times 10^{-31}$ kg
mass proton	$m_p = 1.672\,621\,637(83) \times 10^{-27}$ kg
	$= 1.007\,276\,466\,77(10)$ u
m_e/m_p	$= 5.446\,170\,2177(24) \times 10^{-4}$
atomic mass unit	$u = 1.660\,538\,782(83) \times 10^{-27}$ kg
Avogadro number	$N_A = 6.022\,141\,79(30) \times 10^{23}$ mol^{-1}
Boltzmann constant	$k = 1.380\,6504(24) \times 10^{-23}$ J/K
gas constant kN_A	$R = 8.314\,472(15)$ J mol^{-1} K^{-1}
molar volume	$V_m = 22.710\,98(40) \times 10^{-3}$ m^3/mol
ideal gas 273.15 K, 100 kPa	
Faraday eN_A	$F = 96\,485.3399(24)$ C/mol
Bohr radius	$a_0 = 5.291\,772\,0859(36) \times 10^{-11}$ m
$a_0 = \hbar/(m_e c\alpha) = 10^7\,(\hbar/ce)^2/m_e$	
Bohr magneton	$\mu_B = 9.274\,009\,15(23) \times 10^{-24}$ J/T
$\mu_B = e\hbar/2m_e$	
nuclear magneton	$\mu_N = 5.050\,783\,24(13) \times 10^{-27}$ J/T
magnetic moment electron	$\mu_e = -9.284\,763\,77(23) \times 10^{-24}$ J/T
magnetic moment proton	$\mu_p = 1.410\,606\,662(37) \times 10^{-26}$ J/T
g-factor electron	$g_e = -2.002\,319\,304\,3622(15)$
g-factor proton	$g_p = 5.585\,694\,713(46)$
fine structure constant	$\alpha = 7.297\,352\,5376(50) \times 10^{-3}$
$\alpha^{-1} = 4\pi\varepsilon_0\hbar c/e^2$	$\alpha^{-1} = 137.035\,999\,679(94)$
proton gyromagnetic	$\gamma_p = 2.675\,222\,099(70) \times 10^8$ s^{-1}T^{-1}
ratio	$\gamma_p/2\pi = 42.577\,4821(11)$ MHz/T

Physical constants

conductance quantum $\quad\quad G_0 = 7.748\,091\,7004(53) \times 10^{-5}$ S

Josephson constant $\quad\quad\quad K_J = 4.835\,978\,91(12) \times 10^{14}$ Hz/V

magnetic flux quantum $\quad\quad \Phi_0 = 2.067\,833\,667(52) \times 10^{-15}$ Wb

$\quad G_0 = 2e^2/h;\ K_J = 2e/h;\ \Phi_0 = h/2e$

Stefan–Boltzmann constant $\quad\quad \sigma = 5.670\,400(40) \times 10^{-8}$

$\quad \pi^2 k^4/(60\hbar^3 c^2);\ U = \sigma T^4$ (black body radiation) $\quad$ W m^{-2}K^{-4}

Rydberg constant $\quad\quad\quad R_\infty = 10\,973\,731{,}568\,527(73)$ m^{-1}

$\quad \alpha^2 m_e c/2h$

masses of neutron (n), deuteron (d) and muon (μ)

n:	$1.674\,927\,211(84) \times 10^{-27}$ kg	$= 1.008\,664\,915\,97(43)$ u
d:	$3.343\,583\,20(17) \times 10^{-27}$ kg	$= 2.013\,553\,212\,724(78)$ u
μ:	$1.883\,531\,30(11) \times 10^{-28}$ kg	$= 0.113\,428\,9256(29)$ u

Relative standard deviations

g_e	7.4×10^{-13}	g_p	8.2×10^{-9}
R_∞	6.6×10^{-12}	e, K_J, Φ_0	2.5×10^{-8}
m_d/u	3.9×10^{-11}	$h, N_A, u, m_e,$	
m_p/u	1.0×10^{-10}	m_p, m_d, m_n	5.0×10^{-8}
$m_e/u, m_n/u,$		k, R, V_m	1.7×10^{-6}
m_e/m_p	4.2×10^{-10}	σ (Stefan-B.)	7.0×10^{-6}
α, a_o, G_0	6.8×10^{-10}	G	1.0×10^{-4}

Accuracies of derived quantities

If y_k is a product of powers of physical constants x_i:

$y_k = a_k \Pi_{i=1}^N x_i^{p_{ki}}$ (a_k is a constant), then

$$\epsilon_k^2 = \sum_{i=1}^N p_{ki}^2 \epsilon_i^2 + 2\sum_{j<i}^N p_{ki} p_{kj} r_{ij} \epsilon_i \epsilon_j,$$

where ϵ_k = relative standard deviation and r_{ij} = correlationcoefficient between i and j (For r: see website)

CODATA 2006 http://physics.nist.gov/cuu/constants/

Probability distributions

Continuous one-dimensional probability functions

x is a real variable from a *domain* $\mathcal{D}$; the *probability density function* (pdf) $p(x)$ is real; $p(x) \geq 0$. $p(x)\,dx$ is the probability of finding a sample X in the interval $(x, x + dx)$.

$p(x)$ is normalized: $\int_{\mathcal{D}} p(x)\,dx = 1$ (if $p(x)$ cannot be normalized, it is called an *improper* pdf).

The value of x for which $p(x)$ is a maximum, is called the *mode*.

The *expectation* or *expected value* of a function $g(x)$ over the pdf $p(x)$ is defined as the functional

$$E[g(x)] \overset{\text{def}}{=} \int_{\mathcal{D}} g(x)p(x)\,dx.$$

mean: $\mu = E[x]$.

variance: $\sigma^2 = E[(x - \mu)^2]$.

standard deviation (std) σ: root of the variance.

n-th moment: $\mu_n \overset{\text{def}}{=} E[x^n]$.

n-th central moment $\mu_n^c \overset{\text{def}}{=} E[(x - \mu)^n]$.

skewness: $E[(x - \mu)^3/\sigma^3]$.

kurtosis: $E[(x - \mu)^4/\sigma^4]$.

excess: *kurtosis*-3.

   characteristic function $\Phi(t)$:

$$\Phi(t) \overset{\text{def}}{=} E[e^{itx}] = \int_{\infty}^{\infty} e^{itx} p(x)\,dx$$

$$= \sum_{n=0}^{\infty} \frac{(it)^n}{n!} E[x^n] = \sum_{n=0}^{\infty} \frac{(it)^n}{n!} \mu_n$$

$\Phi(t)$ generates the moments μ_n. The moments are also given by the *derivatives* of the characteristic function at $t = 0$:

$$\Phi^{(n)}(0) = \frac{d^n \Phi}{dt^n}\Big|_{t=0} = i^n \mu_n.$$

Special case: $\mu_2 = \sigma^2 + \mu^2 = -(d^2 \Phi(t)/dt^2)_{x=0}$.

Probability distributions

(one-dimensional functions – continued)
Cumulative distribution function (cdf) $P(x)$:

$$P(x) \stackrel{\text{def}}{=} \int_a^x p(x')\, dx',$$

where a is the lower limit of the domain of x (usually $-\infty$). $P(x)$ is a monotonously non-decreasing function of x, starting at 0 and ending at 1. The value of x for which $P(x) = 0.5$ is the *median*; when $P(x) = 0.25, x$ is the *first quartile*; when $P(x) = 0.75, x$ is the *third quartile*; when $P(x) = 0.01n, x$ is the n-th *percentile*.
Survival function (sf): $S(x) = 1 - P(x)$.

Continuous two-dimensional probability functions

Joint pdf: $p(x, y)\, dx\, dy$ is the probability for a sample pair (X, Y) to find the value X in the interval $(x, x + dx)$ *and* the value Y in the interval $(y, y + dy)$. $p(x, y) \geq 0$; $\int p(x, y)\, dx\, dy = 1$.
Conditional pdf: $p(x|y)\, dx$ (p *of x given y*) is the probability for a sample pair (X, Y) that a sample X occurs in the interval $(x, x + dx)$ *while* Y has the value y.
Marginal pdf: $p_x(x) = \int p(x, y)\, dy$ is the probability for a sample pair (X, Y) that a sample X occurs in the interval $(x, x + dx)$ *irrespective* of the value of Y.

$$p(x|y) = p(x, y)/p_y(y),$$
$$p(x, y) = p_x(x)\, p(y|x) = p_y(y)\, p(x|y),$$
$$p(x|y) = p_x(x) \text{ if } x \text{ and } y \text{ are independent,}$$
$$p(x, y) = p_x(x)\, p_y(y) \text{ if } x \text{ and } y \text{ are independent.}$$

Expectation of $g(x, y)$: $E[g(x, y)] = \int dx \int dy\, g(x, y)\, p(x, y)$.
Mean of x: μ_x is the expectation $E[x] = \int dx \int dy\, x\, p(x, y) = \int x\, p_x(x)\, dx$.
Variance of x: $\sigma_x^2 = C_{xx} = E[(x - \mu_x)^2]$.
Covariance of x and y: $C_{xy} = E[(x - \mu_x)(y - \mu_y)] = \int dx \int dy(x - \mu_x)(y - \mu_y)p(x, y)$.
Correlation coefficient between x and y: $\rho_{xy} = C_{xy}/(\sigma_x \sigma_y)$.
$\mathbf{C} = E[\mathbf{x}\mathbf{x}^{\mathrm{T}}]$ is the *correlation matrix* ($\mathbf{x}$ is the column vector of deviations from the mean).

Student's t-distribution

Student's t-distribution

Let X be a normally distributed variable with expectation 0 and variance σ^2 and Y^2/σ^2 an independent chi-squared distributed variable with ν degrees of freedom. Then $t = \frac{X\sqrt{\nu}}{Y}$ is distributed according to a Student's t-distribution $f(t|\nu)$ with ν degrees of freedom, independent of σ:

$$f(t|\nu)\,dt = \frac{1}{\sqrt{\nu\pi}} \frac{\Gamma[(\nu+1)/2]}{\Gamma(\nu/2)} \left(1 + \frac{t^2}{\nu}\right)^{-(\nu+1)/2} dt.$$

Application: accuracy of the mean

Let $x_1, \ldots, x_n$ be n independent samples from a normal distribution with unknown expectation μ and unknown variance σ^2; let $\langle x \rangle = \frac{1}{n}\sum_{i=1}^{n} x_i$; $S = \sum_{i=1}^{n}(x_i - \langle x \rangle)^2$ and $\hat{\sigma} = \sqrt{S/(n-1)}$, then $t = [(\langle x \rangle - \mu)\sqrt{n}]/\hat{\sigma}$ is distributed according to a Student's t-distribution with $\nu = n - 1$ degrees of freedom. The best estimate for σ is $\hat{\sigma}$. If σ is known, then $\langle x \rangle$ is distributed normally with mean μ and variance σ^2/n. In the latter case $\chi^2 = S/\sigma^2$ satisfies a chi-squared distribution with $\nu = n - 1$ degrees of freedom.

Properties and moments

f is symmetric: $f(-t) = f(t)$; mean = 0
variance $\sigma^2 = \nu/(\nu - 2)$ $(\nu > 2)$; 'skewness' $\gamma_1 = 0$
"excess" $\gamma_2 = E\{t^4\}/\sigma^4 - 3 = 6/(\nu - 4)$
$\lim_{\nu \to \infty} f(t|\nu) = (1/\sqrt{2\pi}) \exp(-t^2/2)$

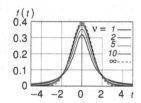

Cumulative distribution

$F(t|\nu) = \int_{-\infty}^{t} f(t'|\nu)\,dt'$
$F(-t|\nu) = 1 - F(t|\nu)$

see table p. 2

Student's t-distribution

Values of t at 75%, 90%, 95%, 99%, and 99.5%

A = acceptance level for two-sided interval $(-t, t)$

$F(t) =$	0.75	0.90	0.95	0.99	0.995
$F(-t) =$	0.25	0.10	0.05	0.01	0.005
$A(\%)$	50	80	90	98	99
$\nu = 1$	1.000	3.078	6.314	31.821	63.657
2	0.816	1.886	2.920	6.965	9.925
3	0.765	1.638	2.353	4.541	5.841
4	0.741	1.533	2.132	3.747	4.604
5	0.727	1.467	2.015	3.365	4.032
6	0.718	1.440	1.943	3.143	3.707
7	0.711	1.415	1.895	2.998	3.499
8	0.706	1.397	1.860	2.896	3.355
9	0.703	1.383	1.833	2.821	3.250
10	0.700	1.372	1.812	2.764	1.169
11	0.697	1.363	1.796	2.718	3.106
12	0.695	1.356	1.782	2.681	3.055
13	0.694	1.350	1.771	2.650	3.012
14	0.692	1.345	1.761	2.624	2.977
15	0.691	1.341	1.753	2.602	2.947
20	0.687	1.325	1.725	2.528	2.845
25	0.684	1.316	1.708	2.485	2.787
30	0.683	1.310	1.697	2.457	2.750
40	0.681	1.303	1.684	2.423	2.704
50	0.679	1.299	1.676	2.403	2.678
60	0.697	1.296	1.671	2.390	2.660
70	0.678	1.294	1.667	2.381	2.648
80	0.678	1.292	1.664	2.374	2.639
100	0.677	1.290	1.660	2.364	2.626
∞	0.674	1.282	1.645	2.326	2.576

Units

Definitions SI basic units

SI: *Système International d'Unités*

Source http://physics.nist.gov/cuu/Units

length: **meter** (m) length of the path traveled by light in vacuum during 1/299 792 458 second (1983).

mass: **kilogram** (kg) mass of the international prototype of the kilogram (1901).

time: **second** (s) duration of 9 192 631 770 periods of the transition between two hyperfine levels of the ground state of the cesium-133 atom (1967).

current: **ampere** (A) current in two infinitely long and thin straight parallel conductors, placed 1 meter apart in vacuum, that exert a force on each other of 2×10^{-7} newton per meter length (1948).

thermodynamic temperature: **kelvin** (K) fraction 1/273.16 of the thermodynamic temperature of the triple point of water (1967).

amount of substance: **mol** (mol) amount of substance which contains as many elementary entities as there are atoms in 0.012 kg of carbon 12. The entities (atoms, molecules, ions, electrons, etc.) must be specified (1971).

luminous intensity: **candela** (cd) radiant intensity of a source that emits monochromatic radiation of frequency 540×10^{12} Hz in a given direction, with intensity of 1/683 W/sr (watt per steradian) (1979).

10^{-1}	deci	d	10^{-2}	centi	c	10^{-3}	milli	m
10^{-6}	micro	μ	10^{-9}	nano	n	10^{-12}	pico	p
10^{-15}	femto	f	10^{-18}	atto	a	10^{-21}	zepto	z
10^{-24}	yocto	y						
10^{1}	deca	da	10^{2}	hecto	h	10^{3}	kilo	k
10^{6}	mega	M	10^{9}	giga	G	10^{12}	Tera	T
10^{15}	peta	P	10^{18}	exa	E	10^{21}	zetta	Z
10^{24}	yotta	Y						

Units

Derived SI units

plane angle (circle: 2π)	$\alpha, \ldots$	**radian**	rad
solid angle (sphere: 4π)	ω, Ω	**steradian**	sr
area	A, S		m^2
volume	V		m^3
frequency	ν	**hertz**	$Hz = s^{-1}$
linear momentum	p		$kg\,m\,s^{-1}$
angular momentum	L, J		$kg\,m^2 s^{-1}$
specific mass	ρ		kg/m^3
moment of inertia	I		$kg\,m^2$
force	F	**newton**	$N = kg\,m\,s^{-2}$
torque	M		$N\,m$
pressure	p, P	**pascal**	$Pa = N/m^2$
viscosity	η		$N\,s\,m^{-2} =$ $kg\,m^{-1}s^{-1}$
energy	E, w	**joule**	$J = N\,m =$ $kg\,m^2 s^{-2}$
power	P	**watt**	$W = J/s$
charge	q, Q	**coulomb**	$C = A\,s$
electric potential	V, Φ	**volt**	$V = J/C$
electric field	E		V/m
dielectric displacement	D		C/m^2
capacity	C	**farad**	$F = C/V$
resistance	R	**ohm**	$\Omega = V/A$
specific resistance	ρ		$\Omega\,m$
conductance	G	**siemens**	$S = \Omega^{-1}$
specific conductance	σ, κ		S/m
inductance	L	**henry**	$H = Wb/A$
magnetic flux	Φ	**weber**	$Wb = V\,s$
magnetic field	H		A/m
magnetic flux density	B	**tesla**	$T = Wb/m^2$
luminous flux	Φ	**lumen**	$lm = cd.sr$
illuminance	I	**lux**	$lx = lm/m^2$
activity (radionuclide)	A	**becquerel**	$Bq = s^{-1}$
absorbed dosis	D	**gray**	$Gy = J/kg$
dose equivalent	H	**sievert**	$Sv = J/kg$

Units

Non-SI units (incl. British, US) (see also **atomic units** on p. 5)

length: **fermi** (fm) $= 10^{-15}$ m; **Ångstrom** (Å) $= 10^{-10}$ m; **mil** (mil) $=$ 0.001 in; **inch** (in) $= 2.54$ cm (exact); **foot** (ft) $= 12$ in $= 0.304\,8$ m; **yard** (yd) $= 3$ ft $= 0.914\,4$ m; **fathom** $= 6$ ft $= 1.828\,8$ m; **cable** $= 720$ ft $=$ 185.2 m; **(statute) mile** $= 1760$ yd $= 1609.34$ m; **nautical mile** (nm) $=$ 1852 m; **astronomical unit** (AU) $= 1.495\,978\,70 \times 10^{11}$ m; **light year** (Ly) $= 9.4605 \times 10^{15}$ m; **parsec** (pc) $= 3.086 \times 10^{16}$ m.

area: **barn** (b) $= 10^{-28}$ m^2 $= 100$ fm^2; **are** (a) $= 100$ m^2; **hectare** (ha) $=$ 10^4 m^2; **acre** $= 4840$ sq. yd $= 4046.87$ m^2; **sq. mile** $= 640$ acres $= 2.59$ km^2.

volume: **Br. fluid ounce** fl oz) $= 28.41$ cm^3; **US fl. oz** $= 29.572\,9$ cm^3; **US liq. pint** $= 16$ US fl. oz $= 473.2$ cm^3; **Br. pint** (pt) $= 20$ Br. fl. oz $=$ 568.2 cm^3; **US liq. quart** $= 2$ US liq. pt $= 946.3$ cm^3; **liter** (l) $= 1$ dm^3; **Br. quart** (qt) $= 2$ Br. pt $= 1.136$ dm^3; **US gallon** $= 4$ US liq. qt $= 231$ in^3 $= 3.785\,4$ dm^3; **(Br.) imperial gallon** (gal) $= 4$ Br. qt $= 4.546$ dm^3; **bushel** $= 8$ imp. gal; **barrel** $= 42$ US gal.; **ton** $= 1$ m^3; **register ton** $= 100$ ft^3 $=$ 2.83 m^3.

mass: **u** (unified atomic mass unit) $= 1.660\,538\,782(83) \times 10^{-27}$ kg; **grain avdp** (gr) $= 64.79891$ mg (exact); **(Br.) drachme** $=$ **(US) dram** $=$ 60 gr $= 3.887\,934\,6$ g; **ounce avdp** (oz) $= 28.349\,527$ g (exact); **troy ounce (apothecary ounce)** $= 480$ gr $= 31.103\,4768$ g; **pound avoirdupois** (lb) $=$ 16 oz $= 7000$ grain $= 0.453\,592\,37$ kg (exact); **(Br.) stone** $= 14$ lbs $= 6.35$ kg; **ton** $= 1000$ kg.

time: **minute** (min) $= 60$ s; **hour** (h) $= 3600$ s.

temperature: t **degree Celsius** (°C) $= t + 273.15$ K; f **degree Fahrenheit** (°F) $= (f - 32) \times 5/9$ °C.

velocity: **knot** $=$ nautical mile/h $= 0.514\,44$ m/s.

force: **dyne** (dyn) $= 10^{-5}$ N; **poundforce** (lbf) $= 4.448\,22$ N; **kilogramforce** (kgf) $= 9.806\,65$ N (exact).

Units

(non SI units, *continued*)

pressure: **mm Hg** (torr) $= 101\,325/760$ Pa (exact) $= 133.322$ Pa;
pound per sq. inch (psi) $= 6\,894.76$ Pa; **technical atmosphere** (at) $=$
kgf/cm^2 $= 98\,066.5$ Pa (exact); **bar** (bar) $= 10^5$ Pa; **normal atmosphere**
(atm) $= 101\,325$ Pa (exact).

energy: **hartree** $(E_h) = 4.359\,743\,94(22) \times 10^{-18}$ J; **erg** (erg) $= 10^{-7}$ J;
thermochemical calorie (cal$_{th}$) $= 4.184$ J; **15° calorie** (cal$_{15}$) $= 4.1855$ J;
Int. Table calorie (cal$_{IT}$) $= 4.1868$ J; **Br. thermal unit** (Btu) $= 1055.87$ J;
kilowatt hour (kWh) $= 3.6$ MJ; **ton coal equiv.** (tse) $= 29.3$ GJ; **ton oil**
equiv. (toe) $= 45.4$ GJ; m^3 **natural gas** (average, 0 °C, 1 atm) $= 39.4$ MJ.

power: **horsepower** (metric, PS) $= 75$ kgf m/s $= 735.5$ W; **horsepower**
(mechanical, hp) $= 550$ lbf ft/s $= 745.7$ W.

viscosity: **poise** (p) $=$ g cm^{-1}s^{-1} $= 0.1$ kg m^{-1}s^{-1}; *kinematic viscosity:*
stokes (St) $= 10^{-4}$ m^2/s

(radio)activity, dose: **curie** (Ci) $= 3.7 \times 10^{10}$ Bq; **röntgen** (R) $=$
2.58×10^{-4} C/kg; **rad** (rad, rd) $= 0.01$ Gy; **rem** (rem) $= 0.01$ Sv.

light: **stilb** (sb) $=$ cd/cm^2; **phot** (ph) $=$ cd cm^{-2}sr^{-1}.

electrostatic units (esu): c.g.s. unit of charge
(g$^{1/2}$cm$^{3/2}$s^{-1}), such that $4\pi\varepsilon_0 = 1$ (dimensionless): *charge:* $10^{-9}/$
$2.997\,924\,58$ C; *current:* $10^{-9}/2.99\ldots$ A; *dipole moment:* $10^{-11}/2.99\ldots$ C m;
debye (D) $= 10^{-18}$ esu $= 10^{-29}/2.99\ldots$ C; *el. pot.:* $299.7\ldots$ V; *el. field:*
$2.99\ldots \times 10^4$ V/m.

electromagnetic units (emu): c.g.s. unit of current (g$^{1/2}$cm$^{1/2}$s^{-1}), such
that $\mu_0/4\pi = 1$ (dimensionless): *current:* **abampere** (abamp) $= 10$ A;
magn. field: **oerstedt** (Oe) $= (1/4\pi)$ abamp/cm $= 10^3/4\pi$ A/m; *magn. flux*
density ("induction"): **gauss** (G) $= 10^{-4}$ T; *magn. flux:* **maxwell** (Mx) $=$
10^{-8} Wb.

Units

Atomic units (a.u.)

The basic a.u. are the Bohr radius a_0, the electron mass m_e, Dirac's constant $\hbar$ and the elementary charge e: $m_e = 1$ a.u., $\hbar = 1$ a.u., $c = 1/\alpha$ a.u., $e = 1$ a.u., $4\pi\varepsilon_0 = 1$ a.u.

mass	$m_e = 9.109\,382\,15(45)\times10^{-31}$ kg
length	$a_0 = 5.291\,772\,0859(36)\times10^{-11}$ m
charge	$e = 1.602\,176\,487(40)\times10^{-19}$ C
time	$a_0/(\alpha c) = 2.418\,884\,326\,505(16)\times10^{-17}$ s
	$= (4\pi R_\infty c)^{-1}$
velocity	$\alpha c = 2.187\,691\,2541(15)\times10^{6}$ m/s
energy	$\hbar^2/(m_e a_0^2) = e^2/(4\pi\varepsilon_0 a_0) = \alpha^2 mc^2$
	$= 2R_\infty hc =$
(hartree)	$E_h = 4.359\,743\,94(22)\times10^{-18}$ J
	$= 2\,625.312\,93(13)$ kJ/mol
	$= 627.464\,850(32)$ kcal/mol
	$= 27.211\,383\,86(68)$ eV

Molecular units

This is a consistent system of units for 'molecular' quantities, useful for molecular modeling and simulation. Coulomb forces have an *electric factor* coefficient $f = 1/(4\pi\varepsilon_0)$: $F = fq_1 q_2/r^2$ (see table). The unit for f is kJ mol^{-1} nm e^{-2}.

mass	u	$= 1.660\,538\,86(28)\times10^{-27}$ kg
length	nm	$= 10^{-9}$ m
time	ps	$= 10^{-12}$ s
velocity	nm/ps	$= 1000$ m/s
energy	kJ/mol	$= 1.660\,538\,86\times10^{-21}$ J
force	kJ mol^{-1} nm^{-1}	$= 1.660\,538\,86\times10^{-12}$ N
pressure	kJ mol^{-1} nm^{-3}	$= 1.660\,538\,86\times10^{5}$
		$= 16.605\,3886$ bar
charge	e	$= 1.602\,176\,53(14)\times10^{-19}$ C
electric factor	f	$= 138.935\,4574(14)$

Index